# A Systems Approach to Lithium-Ion Battery Management

# Recent Artech House Titles in Power Engineering

Dr. Jianhui Wang, Series Editor

*The Advanced Smart Grid: Edge Power Driving Sustainability,* Andres Carvallo and John Cooper

*Battery Management Systems for Large Lithium Ion Battery Packs,* Davide Andrea

*Battery Power Management for Portable Devices,* Yevgen Barsukov and Jinrong Qian

*Designing Control Loops for Linear and Switching Power Supplies: A Tutorial Guide,* Christophe Basso

*Electric Systems Operations: Evolving to the Modern Grid,* Mani Vadari

*Energy Harvesting for Autonomous Systems,* Stephen Beeby and Neil White

*GIS for Enhanced Electric Utility Performance,* Bill Meehan

*Power Line Communications in Practice,* Xavier Carcelle

*Power System State Estimation,* Mukhtar Ahmad

*A Systems Approach to Lithium-Ion Battery Management,* Phillip Weicker

*Synergies for Sustainable Energy,* Elvin Yüzügüllü

# A Systems Approach to Lithium-Ion Battery Management

Phillip Weicker

ARTECH
HOUSE
BOSTON | LONDON
artechhouse.com

Library of Congress Cataloging-in-Publication Data
A catalog record for this book is available from the U.S. Library of Congress.

British Library Cataloguing in Publication Data
A catalogue record for this book is available from the British Library.

Cover design by Vicki Kane

ISBN 13: 978-1-60807-659-8

© 2014 ARTECH HOUSE
685 Canton Street
Norwood, MA 02062

All rights reserved. Printed and bound in the United States of America. No part of this book may be reproduced or utilized in any form or by any means, electronic or mechanical, including photocopying, recording, or by any information storage and retrieval system, without permission in writing from the publisher.
 All terms mentioned in this book that are known to be trademarks or service marks have been appropriately capitalized. Artech House cannot attest to the accuracy of this information. Use of a term in this book should not be regarded as affecting the validity of any trademark or service mark.

10 9 8 7 6 5 4 3 2 1

*To the memory of my mother*

# Contents

| 1 | **Introduction** | **17** |
|---|---|---|
| 1.1 | Battery Management Systems and Applications | 17 |
| 1.2 | State of the Art | 18 |
| 1.3 | Challenges | 22 |
| **2** | **Lithium-Ion Battery Fundamentals** | **25** |
| 2.1 | Battery Operation | 25 |
| 2.2 | Battery Construction | 26 |
| 2.3 | Battery Chemistry | 29 |
| 2.4 | Safety | 35 |
| 2.5 | Longevity | 38 |
| 2.6 | Performance | 39 |
| 2.7 | Integration | 40 |
| **3** | **Large-Format Systems** | **43** |
| 3.1 | Definition | 43 |

| | | |
|---|---|---|
| 3.2 | Balance of Plant | 45 |
| 3.3 | Load Interface | 46 |
| 3.4 | Variation and Divergence | 47 |
| 3.5 | Application Parameters | 48 |
| **4** | **System Description** | **51** |
| 4.1 | Typical Inputs | 52 |
| 4.2 | Typical Outputs | 54 |
| 4.3 | Typical Functions | 56 |
| 4.4 | Summary | 57 |
| **5** | **Architectures** | **59** |
| 5.1 | Monolithic | 59 |
| 5.2 | Distributed | 61 |
| 5.3 | Semi-Distributed | 61 |
| 5.4 | Connection Methods | 63 |
| 5.5 | Additional Scalability | 65 |
| 5.6 | Battery Pack Architectures | 66 |
| 5.7 | Power Supply | 67 |
| 5.8 | Control Power | 68 |
| 5.9 | Computing Architecture | 69 |
| **6** | **Measurement** | **71** |
| 6.1 | Cell Voltage Measurement | 71 |
| 6.2 | Current Measurement | 77 |
| 6.2.1 | Current Sensors | 78 |
| 6.2.2 | Current Sense Measurement | 84 |
| 6.3 | Synchronization of Current and Voltage | 86 |

| | | |
|---|---|---|
| 6.4 | Temperature Measurement | 87 |
| 6.5 | Measurement Uncertainty and Battery Management System Performance | 92 |
| 6.6 | Interlock Status | 92 |

## 7 Control — 95

| | | |
|---|---|---|
| 7.1 | Contactor Control | 95 |
| 7.2 | Soft Start or Precharge Circuits | 97 |
| 7.3 | Control Topologies | 99 |
| 7.4 | Contactor Opening Transients | 101 |
| 7.5 | Chatter Detection | 102 |
| 7.6 | Economizers | 104 |
| 7.7 | Contactor Topologies | 105 |
| 7.8 | Contactor Fault Detection | 106 |

## 8 Battery Management System Functionality — 111

| | | |
|---|---|---|
| 8.1 | Charging Strategies | 111 |
| 8.1.1 | CC/CV Charging Method | 111 |
| 8.1.2 | Target Voltage Method | 112 |
| 8.1.3 | Constant Current Method | 113 |
| 8.2 | Thermal Management | 114 |
| 8.3 | Operational Modes | 115 |

## 9 High-Voltage Electronics Fundamentals — 119

| | | |
|---|---|---|
| 9.1 | High-Voltage DC Hazards | 119 |
| 9.2 | Safety of High-Voltage Electronics | 120 |
| 9.3 | Conductive Anodic Filaments | 123 |
| 9.4 | Floating Measurements | 124 |
| 9.4.1 | Y-Capacitance | 125 |

| | | |
|---|---|---|
| 9.5 | HV Isolation | 125 |
| 9.6 | ESD Suppression on Isolated Devices | 128 |
| 9.7 | Isolation Detection | 130 |

| **10** | **Communications** | **133** |
|---|---|---|
| 10.1 | Overview | 133 |
| 10.2 | Network Technologies | 133 |
| 10.2.1 | I²C/SPI | 134 |
| 10.2.2 | RS-232 and RS-485 | 134 |
| 10.2.3 | Local Interconnect Network | 136 |
| 10.2.4 | CAN | 136 |
| 10.2.5 | Ethernet and TCP/IP | 137 |
| 10.2.6 | Modbus | 138 |
| 10.2.7 | FlexRay | 138 |
| 10.3 | Network Design | 138 |

| **11** | **Battery Models** | **145** |
|---|---|---|
| 11.1 | Overview | 145 |
| 11.2 | Thévenin Equivalent Circuit | 146 |
| 11.3 | Hysteresis | 151 |
| 11.4 | Coulombic Efficiency | 153 |
| 11.5 | Nonlinear Elements | 154 |
| 11.6 | Self-Discharge Modeling | 157 |
| 11.7 | Physics-Based Battery Models | 158 |
| 11.7.1 | Doyle-Fuller-Newman Model | 158 |
| 11.7.2 | Single Particle Model | 158 |
| 11.8 | State-Space Representations of Battery Models | 161 |
| | References | 163 |

| **12** | **Parameter Identification** | **165** |
|---|---|---|
| 12.1 | Brute-Force Approach | 165 |

| | | |
|---|---|---|
| 12.2 | Online Parameter Identification | 166 |
| 12.3 | SOC/OCV Characterization | 167 |
| 12.4 | Kalman Filtering | 168 |
| 12.5 | Recursive Least Squares | 168 |
| 12.6 | Electrochemical Impedance Spectroscopy | 169 |

## 13 Limit Algorithms 171

| | | |
|---|---|---|
| 13.1 | Purpose | 171 |
| 13.2 | Goals | 172 |
| 13.3 | Limit Strategy | 172 |
| 13.4 | Determining Safe Operating Area | 173 |
| 13.5 | Temperature | 174 |
| 13.6 | SOC/DOD | 177 |
| 13.7 | Cell Voltage | 179 |
| 13.8 | Faults | 180 |
| 13.9 | First-Order Predictive Power Limit | 180 |
| 13.10 | Polarization-Dependent Limit | 181 |
| 13.11 | Limit Violation Detection | 181 |
| 13.12 | Limits with Multiple Parallel Strings | 182 |

## 14 Charge Balancing 183

| | | |
|---|---|---|
| 14.1 | Balancing Strategies | 184 |
| 14.2 | Balancing Optimization | 185 |
| 14.3 | Charge Transfer Balancing | 187 |
| 14.3.1 | Flying Capacitor | 188 |
| 14.3.2 | Inductive Charge Transfer Balancing | 190 |
| 14.3.3 | Transformer Charge Balancing | 193 |

| 14.4 | Dissipative Balancing | 193 |
|---|---|---|
| 14.5 | Balancing Faults | 197 |

## 15 State-of-Charge Estimation Algorithms 199

| 15.1 | Overview | 199 |
|---|---|---|
| 15.2 | Challenges | 199 |
| 15.3 | Definitions | 201 |
| 15.4 | Coulomb Counting | 203 |
| 15.5 | SOC Corrections | 204 |
| 15.6 | OCV Measurements | 205 |
| 15.7 | Temperature Compensation | 206 |
| 15.8 | Kalman Filtering | 206 |
| 15.9 | Other Observer Methods | 211 |
|  | Reference | 212 |

## 16 State-of-Health Estimation Algorithms 213

| 16.1 | State of Health | 213 |
|---|---|---|
| 16.2 | Mechanisms of Failure | 215 |
| 16.3 | Predictive SOH Models | 216 |
| 16.4 | Impedance Detection | 219 |
| 16.4.1 | Passive Methods | 219 |
| 16.4.2 | Active Methods | 221 |
| 16.5 | Capacity Estimation | 223 |
| 16.6 | Self-Discharge Detection | 226 |
| 16.7 | Parameter Estimation | 226 |
| 16.8 | Dual-Loop System | 226 |
| 16.9 | Remaining Useful Life Estimation | 227 |

| | | |
|---|---|---|
| 16.10 | Particle Filters | 227 |
| | Reference | 229 |

## 17 Fault Detection — 231

| | | |
|---|---|---|
| 17.1 | Overview | 231 |
| 17.2 | Failure Detection | 231 |
| 17.2.1 | Overcharge/Overvoltage | 231 |
| 17.2.2 | Over-Temperature | 235 |
| 17.2.3 | Overcurrent | 235 |
| 17.2.4 | Battery Imbalance/Excessive Self-Discharge | 236 |
| 17.2.5 | Internal Short Circuit Detection | 237 |
| 17.2.6 | Detection of Lithium Plating | 237 |
| 17.2.7 | Venting Detection | 237 |
| 17.2.8 | Excessive Capacity Loss | 238 |
| 17.3 | Reaction Strategies | 238 |
| | References | 239 |

## 18 Hardware Implementation — 241

| | | |
|---|---|---|
| 18.1 | Packaging and Product Development | 241 |
| 18.2 | Battery Management System IC Selection | 242 |
| 18.3 | Component Selection | 248 |
| 18.3.1 | Microprocessor | 248 |
| 18.3.2 | Other Components | 249 |
| 18.4 | Circuit Design | 250 |
| 18.5 | Layout | 252 |
| 18.6 | EMC | 252 |
| 18.7 | Power Supply Architectures | 253 |
| 18.8 | Manufacturing | 254 |

## 19 Software Implementation — 257

| | | |
|---|---|---|
| 19.1 | Safety-Critical Software | 258 |
| 19.2 | Design Goals | 259 |

| | | |
|---|---|---|
| 19.3 | Analysis of Safety-Critical Software | 259 |
| 19.4 | Validation and Coverage | 260 |
| 19.5 | Model Implementation | 262 |
| 19.6 | Balancing | 263 |
| 19.7 | Temperature Impact on State of Charge Estimation | 264 |
| **20** | **Safety** | **265** |
| 20.1 | Functional Safety | 265 |
| 20.2 | Hazard Analysis | 265 |
| 20.3 | Safety Goals | 269 |
| 20.4 | Safety Concepts and Strategies | 270 |
| 20.5 | Reference Design for Safety | 270 |
| **21** | **Data Collection** | **275** |
| 21.1 | Lifetime Data Gathering | 275 |
| **22** | **Robustness and Reliability** | **279** |
| 22.1 | Failure Mode Analysis | 280 |
| 22.2 | Environmental Durability | 283 |
| 22.3 | Abuse Conditions | 285 |
| 22.4 | Reliability Engineering | 286 |
| **23** | **Best Practice** | **287** |
| 23.1 | Engineering System Development | 287 |
| 23.2 | Industry Standards | 288 |
| 23.3 | Quality | 289 |

| | | |
|---|---|---|
| **24** | **Future Developments** | **291** |
| 24.1 | Subcell Modeling | 291 |
| 24.2 | Adaptive Algorithms | 291 |
| 24.3 | Advanced Safety | 292 |
| 24.4 | System Integration | 292 |
| | **Endnotes** | **293** |
| | **About the Author** | **295** |
| | **Index** | **297** |

# 1

# Introduction

## 1.1 Battery Management Systems and Applications

The advent of lithium-ion batteries has brought a significant shift in the area of large-format battery systems. Previously limited to heavy and bulky lead-acid storage batteries, large-format batteries were used only where absolutely necessary as a means of energy storage. The improved energy density, cycle life, power capability, and durability of lithium-ion cells has given us electric and hybrid vehicles with meaningful driving range and performance, grid-tied energy storage systems for integration of renewable energy and load leveling, backup power systems, and other innovations. Future uses will include not only better versions of the above, but also other electrified means of transportation including marine and aviation, as well as enabling further decentralization and distribution of energy generation and storage. Alongside all of the innovation that has taken place in semiconductor, software, and microprocessor technology in the past 30 years, the appearance of the lithium-ion battery has quietly enabled the generation of portable electronic devices that are now commonplace in all societies and given us the connected and mobile world we are used to. In the future, this battery technology will give us new ways to generate, use, and store energy and free us from the perils of nonrenewable energy sources.

However, with any source of stored energy, effective management is required to ensure that an uncontrolled release of that energy does not occur, and large-format lithium-ion battery systems are no exception. No form of stored energy can be completely without danger. Although safety and reliability of these cells improves continuously, so does the density of the stored energy and the power capability, making for new management challenges with every new generation of battery cells.

Battery systems must be protected from a variety of situations in which they could become hazardous. The complete system must account for all of the required forms of protection (including controls, mechanical, thermal, and environmental), but large battery systems, such as those used for grid-tied storage or electrified vehicles, incorporate sophisticated electronics and software that work together to measure battery parameters, determine the battery's condition, and control the system to ensure that it operates as desired. This electronic system is referred to as a battery management system.

## 1.2 State of the Art

Battery systems are expected to be a safe and reliable source of energy that delivers the high performance that modern battery cells and chemistries have to offer. Modern battery management system components offer the high reliability and quality that are expected of electronic control systems in nearly every field of application.

Lithium-ion batteries offer the promise of new ways to store and use energy but are only just beginning to see applications beyond a small number of cells to power low-voltage, low-power devices. Since 2010, a number of hybrid and electric vehicles (powered by lithium-ion battery systems, such as the Chevrolet Volt, shown in Figure 1.1) have been introduced to the market, storing multiple kilowatt-hours of energy and operating at hundreds of volts. The recent advances in smart-grid technologies for applications like integrating renewable generation and maintaining power quality have also created increasing demand for stationary energy storage products from building scale (tens of kilowatts) to utility scale (megawatts). As battery costs fall and performance improves (currently, metrics such as energy density are improving at approximately 5% to 6% per year), adoption of large-format lithium-ion technology will replace other types of batteries used in applications such as aviation, rail, and marine. Each of these applications brings different expectations, requirements, and standards. The need for effective battery management will grow accordingly with each new application and advance in battery technology.

Recent stories have highlighted the need for great care in the development of large-format lithium-ion batteries. A series of notebook computer fires brought to light the danger of a malfunction of even small groups of cells in 2006 to 2008. The introduction of electric vehicles powered by lithium-ion batteries was punctuated by thermal events that occurred during crash testing and on-road accidents. Early adoption of lithium-ion batteries for aviation led to a number of in-flight events and caused a high-profile grounding

**Figure 1.1** Chevrolet Volt, a plug-in hybrid electric vehicle (PHEV) using large-format lithium-ion battery. (Courtesy General Moters Company.)

of all Boeing 787 aircraft only months after its introduction (see Figure 1.2). Lithium-ion battery chemistries are much less tolerant to abusive conditions such as overcharging, over-discharging, over-temperature, and excessive current

**Figure 1.2** APU Battery from Boeing 787, damaged by thermal event. (Courtesy National Transportation Safety Board.)

than other types of batteries. High-voltage systems always carry a risk of electric shock as well as the thermal risks associated with battery systems.

Despite the widely-used, general designation of lithium-ion batteries, there are actually a wide variety of materials and electrochemistries used, each with significant implications for performance, lifetime, and safety. This selection of materials has a significant impact on the needs of battery management system and further broadens the challenge of battery management system development.

As a battery management system is responsible for battery safety, the software and hardware as well as the integration and testing of these two together should be developed in accordance with best practices for safety-critical systems. The attention to safety should extend throughout the entire development and life cycle of the battery management system product, including applications, design, implementation, testing, deployment, and service.

Up until recently, battery management system development required a great deal of specialized analog circuitry using a large number of discrete devices, or a cumbersome repurposing of devices originally intended for small lithium-ion battery packs as found in laptop computers and other devices. Custom devices attempted to reduce the complexity of systems, but low numbers of lithium-ion systems did not allow for applications beyond limited volumes. In response to the increased interest in large-format systems, in 2008, a number of specialized integrated circuits were introduced with the large-format battery management system market in mind. In contrast to previous devices that were not intended for applications requiring isolation and high voltages, these integrated coils (ICs) were targeted specifically at the applications discussed above. These have allowed for more compact electronics with fewer components leading to an easier implementation and generally higher reliability. Second-generation versions of these devices began to appear in 2011 and continue to support higher levels of safety, redundancy, and flexibility. Despite the availability of compact and reliable high-voltage analog circuitry, the integration of these devices into a fully operational system requires significant care and analysis to ensure that high-reliability and safety-critical systems can be deployed with confidence.

Large-scale deployment of high-energy lithium-ion battery systems has led to the creation of a number of safety standards for their design, manufacture, and validation. At this time, there are few standards and guidelines that are universally accepted across all applications. Battery management system and battery system designers are often forced to reduce their engineering challenges to the first principles and undertake original development in their work to solve battery management problems. The high voltages and high energy content involved are outside the normal realm of many who are experienced in the development of embedded control systems.

There is certainly no ideal solution for all battery management needs. Numerous important choices must be made, considering the types of applications for the battery system, the choice of battery technology, the product mix with which the battery management system is expected to work, the needs for auxiliary components, and the degree of scalability and modularity needed.

Lithium-ion batteries present unique challenges for calculating the battery state of charge requiring more complex methods than other batteries, and many of the applications in which large-format lithium-ion systems are used require higher accuracy than many previous battery-powered applications. A number of simulation and modeling techniques have been proposed in scientific literature for accomplishing many of the complex software tasks required of a battery management system such as computing state of charge, battery capacity, model parameters, and power limits. Many concepts from control theory and state estimation have been used for the determination of battery parameters. This book will focus on a review of the most common techniques and discuss suitable practical implementations that can be used for the development of real-world systems.

A large-format battery system contains numerous components besides the battery cells themselves, including sensors and actuators with which the battery management system often interfaces. These components (often referred to as the *balance of plant*) work together with the monitoring circuits and processing algorithms to provide complete control and to maximize battery performance.

A good battery management system is necessary but not sufficient to ensure a safe battery system. Battery safety is a holistic concept involving all aspects layers from active materials and cell ingredients, cell design, module construction, battery pack mechanical, electrical and thermal design, and management and control devices and protection schemes.

Special consideration is given to the robustness that is required against numerous failure modes of both the battery cells and the battery management system itself. Battery systems have sensitivities to excesses of temperature, charge, and current, as well as possible adverse reactions to mechanical abuse or manufacturing defects. Battery management system failures can cause these events to go unnoticed, or worse, exacerbate their symptoms. Battery management system designers face the considerable responsibility of ensuring that large-format battery systems are protected from misuse and the risks of battery defects are mitigated. Concepts for reliability and safety analysis, redundant architectures, and risk quantification are emerging. System safety is a foremost consideration that affects every component of a well-designed battery management system as well as their integration and the processes used to design, build, and test the final product, and detailed analysis techniques are used to quantify the risks associated with the final integrated battery system and battery management system.

Modern battery management systems are capable of performing all or nearly all of their required functions simply by measuring battery responses while the battery operates in the normal manner. Previously, battery management systems may have required special cycling or current injection to execute the required measurements, but this practice has become less common.

An appropriate battery management system does not significantly increase the cost of the battery system. The methods, designs, and techniques in this book are intended to lead to systems that provide acceptable performance at a reasonable cost. Often, reduction in the complexity of circuits, models, and algorithms is necessary to achieve practical cost targets. Effective ways of making these trade-offs are outlined.

The development of predictive algorithms for measuring and calculating hidden battery parameters such as state of charge has advanced considerably as well. Modern battery management systems incorporate a battery model developed from first principle models of the underlying physics and/or characterization testing performed on the cell. This model can be updated by dynamic measurement of the model parameters while the battery is in operation. This ensures not only that the model remains correct, but also that the battery's deterioration over time can be monitored and the health of the battery can be measured and acted upon as appropriate.

Success in battery management system development requires proficiency in the domains of analog and digital hardware design, software analysis, system engineering, safety analysis, and a working knowledge of electrochemical energy storage fundamentals and the modeling of electrochemical systems.

## 1.3 Challenges

As lithium-ion batteries are incorporated into more advanced devices, the performance of the management electronics will need to evolve accordingly. Mission-critical applications such as aviation, aerospace, medical, and defense devices will require higher levels of reliability and eliminate the possibility of shutting down the battery system as an option to ensure system safety. Additionally, in these types of applications, errors in the complex problem of battery state prediction are not tolerable in many of these critical applications.

New battery technologies will also create a need for new management capabilities. New active materials are continuously developed for improved safety, energy capacity, and life, which create different battery cell behavior. These additional effects will need to be accounted for in battery management systems to ensure that the additional safety, performance, and reliability offered by these technologies are fully realized.

As development cycles accelerate and the use of modeling and simulation tools to design battery cells becomes more widespread, battery management system development will no longer need to wait for functional battery cells to begin characterization of the battery system. The macroscopic electrical performance can be predicted from the battery cell design and microscale models of the active materials.

As power semiconductors capable of higher operating voltages become available, systems are likely to move to higher voltages, bringing new challenges associated with larger numbers of cells and higher electrical safety hazards.

Battery management systems are currently limited in the amount of degradation or change in cell behavior that can be tolerated while still providing acceptable performance. As the acceptance of battery-powered systems grows, battery systems will remain in service for longer service lives and the expectations for replacement or service intervals will increase. Battery management systems will need to cope with potentially significant changes in battery behavior while maintaining accurate prediction of the battery condition.

The field of lithium-ion batteries is likely to be a fertile area for many years to come, leading to new advances in the fields of transportation and energy storage. With each new wave of battery improvements, battery management systems will need to keep up with the changes. The field of large-format battery systems is only beginning and many new developments await in the area of battery management systems. This exciting field will contribute to changing the landscape of how energy is produced and consumed and lead to a sustainable and ecologically viable future.

# 2

# Lithium-Ion Battery Fundamentals

## 2.1 Battery Operation

Lithium-ion batteries operate by the same electrochemical fundamentals as all batteries (which will not be covered here), but a few important distinctions are important to consider in the development of battery management systems for lithium-ion based systems.

In normal cases, there is no metallic lithium inside a lithium-ion battery cell. This improves significantly the safety of the cell as well as improving the ability of the battery to cycle many times.

In both electrodes, the lithium ions are *intercalated* into the electrode materials. Intercalation is highly reversible, compared to many other electrochemical processes, leading to electrode stability and high cycle life of lithium ion batteries.

Lithium-ion batteries have very low self-discharge rates compared to other types of batteries. As such, they may be used in applications where the batteries are not charged for long periods of time but are expected to deliver good performance when the system is powered on.

Lithium-ion batteries are not susceptible to "memory" effects, meaning that users are free to charge and discharge them much more flexibly than other batteries.

Lithium-ion batteries generally offer a very high coulombic efficiency throughout the state of charge range. Other types of batteries are capable of being trickle-charged continuously at a low rate, even after 100% state of charge (SOC) has been reached, leading to a simple method of determining the fully charged state as well as providing an intrinsic balancing function when the pack is fully charged. Lithium-ion batteries cannot be operated this way—trickle charging at even very low rates will lead to overcharging, battery damage, and possible unsafe conditions.

Most batteries have an aqueous (water-based) electrolyte. The high voltages present in lithium ion cells prevent the use of an aqueous electrolyte (electrolysis begins to occur around 2V). The nonaqueous electrolytes are composed of organic solvents that are flammable and have high vapor pressures. The flammability and high reactivity of these electrolytes pose more severe flammability hazards than other types of batteries.

The high performance of lithium-ion batteries is a double-edged sword. These batteries are often selected for an application based on their high energy content and power capability, but this high performance can lead to a higher severity event if things go wrong. Short-circuit currents can be much higher and an uncontrolled release of energy can be larger. In addition, a number of additional internal reactions that take place during the breakdown of lithium-ion batteries can release additional energy.

## 2.2 Battery Construction

A lithium-ion cell (an example is shown in Figure 2.1) consists of the following components:

- Positive electrode (often referred to as the cathode, although the cathode is strictly speaking the electrode where reduction occurs);
- Negative electrode (often referred to as the anode);
- Electrolyte;
- Separator;
- Enclosure.

Both electrodes consist of electrode material that is coated onto a metal foil that acts as a substrate and current collector. The electrode material contains active material that stores lithium, substances to increase conductivity of both lithium ions and electrons and binders and other materials to provide structural integrity and good adhesion to the metal foil.

The additional substances in the electrode materials provide electronic conductivity between the active material particles and the current collector, as well as ionic conductivity between the electrolyte and the active material.

In cells using a liquid electrolyte, the electrolyte is generally composed of nonaqueous organic solvents containing a dissolved lithium salt. Common materials for electrolytes are ethylene carbonate (EC), diethylene carbonate

# Lithium-Ion Battery Fundamentals

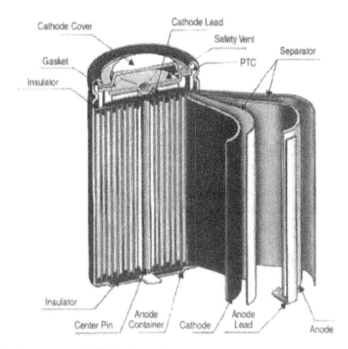

**Figure 2.1** Components of a lithium-ion cell.

(DEC), and dimethylene carbonate (DMC). The lithium salt is often lithium hexafluorophosphate (LiPF$_6$). In contrast to aqueous electrolytes that are often strongly acidic or alkaline solutions, lithium-ion electrolytes are not corrosive but the solvents used are highly flammable and have a relatively high vapor pressure, leading to risks of fire and explosion if cells are vented. Most lithium-ion batteries are not a wet-type battery in which the interior of the enclosure is flooded with electrolyte; most of the electrolyte is absorbed into the active material and separator. Figure 2.2 shows a typical electrolyte, electrode, and active metal structure

Some lithium cells are built with a polymer electrolyte. The use of a conductive polymer avoids the flammability of the liquid electrolyte at the expense of lower ionic conductivity than liquid batteries.

The separator is a porous polymer film used to separate the two electrodes while providing a barrier across which lithium ions can travel. The most common materials used are polyethylene and polypropylene. The extremely thin separator film must effectively keep the anode and cathode apart to prevent short circuits while providing an effective path for lithium ions to travel between electrodes.

The entire cell must be enclosed in a container. The container must be sealed to prevent electrolyte loss and contamination. It must be durable enough

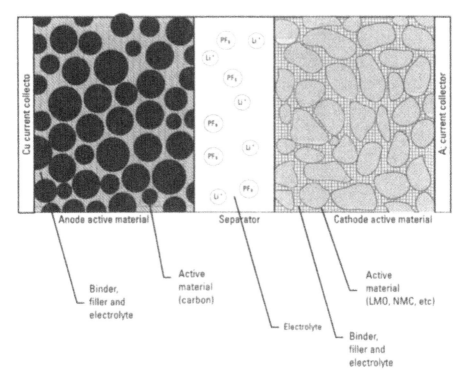

**Figure 2.2** Typical active material structure.

to protect the relatively delicate contents of the cell and offer some resistance to abuse. The most common forms of cell enclosure include:

- *Cylindrical:* At the time of this writing, the most common standard form factors for lithium-ion batteries were small-format (less than 5 Ah) cylindrical cells. The two most common standard sizes are referred to as the 18650 (18-mm diameter and 65 mm in length; see Figure 2.3) and the 26650 (26-mm diameter and 65 mm in length). Usually the majority of the can-shaped enclosure is the negative terminal of the cell with the positive terminal and vent at one end. Cylindrical cells use a single "jelly-roll" of electrodes and separator.

- *Prismatic:* Prismatic cells (see Figure 2.4) offer a robust metallic enclosure that is rectangular in shape. The enclosure is usually sealed via a laser welding process after the assembly but before electrolyte filling. Prismatic cells can be more tightly packed than cylindrical cells (the maximum packing factor is about 90% for cylindrical cells) and there are a wider variety of sizes and shapes available. Vent and terminal locations may vary between manufacturers.

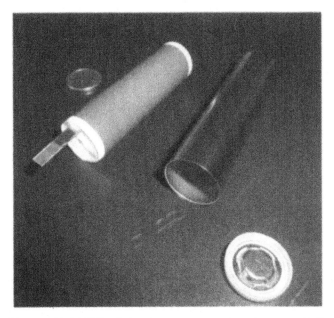

**Figure 2.3** Disassembled 18650 cell showing pressure vent, enclosure, and jelly roll.

- *Pouch:* Pouch cells (see Figure 2.5) are enclosed in a pouch made from two pieces of plastic film fused together around the periphery. Metallic tabs protrude from the pouch for connecting to the cell. Pouch cells are usually very flat in one dimension which is the direction in which the lithium ions travel. The electrode and separator may be folded, stacked, or wound in pouch or prismatic cells. Pouch cells generally vent at an edge where the pouch is sealed.

## 2.3  Battery Chemistry

The selection of the materials used to store lithium in the two electrodes, as well as (to a lesser extent) the composition of the electrolyte is often referred to as the battery chemistry.

The term "battery chemistry" most often refers to the choice of cathode material. Many lithium-ion batteries use a carbon-based anode (exceptions will be discussed). However, it is important to note that the choice of both anode and cathode materials, as well as other substances in both electrodes, has significant impact on battery behavior.

30   A Systems Approach to Lithium-Ion Battery Management

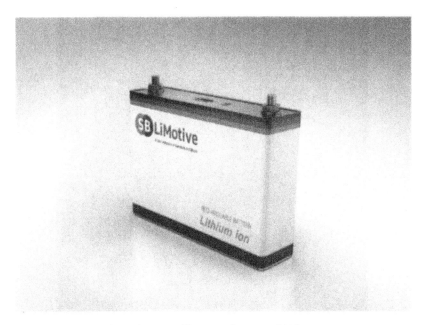

**Figure 2.4**   Prismatic lithium-ion cells. (Courtesy Samsung SDI.)

**Figure 2.5**   Pouch cells and module for an electric vehicle.

From the perspective of a battery management system designer, the choice of battery chemistry will have an impact on the battery management system in a number of ways.

Some common battery chemistry types are listed here:

- *Lithium cobalt oxide (LCO):* Commonly referred to as simply LCO, these cells offer a very high energy density. Most laptop battery packs are made from LCO cells in an 18650 form factor. LCO is used in most of the highest energy density cells currently available. LCO is a toxic material and the potential to release a great deal of energy LCO-based cells have if they are overcharged, or reaches the LCO's degradation temperature. LCO has successfully been used in large-format applications containing a large number of small cells with additional safety features to mitigate the risks of this highly reactive chemistry. LCO is also less stable than other electrode materials and does not offer exceptionally high cycle life performance. As a raw material, LCO has the highest cost per kilogram of common cathode materials, but its high energy density may lead to a lower cost per watt-hour at the cell level than with other materials.
- *Lithium nickel oxide (LNO):* Lithium nickel oxide is a relatively new material for lithium ion battery cathodes. It has even higher energy density than LCO (approximately 15% higher) but has poorer safety characteristics. Nickel-cobalt blends have been developed to take advantage of both the higher energy and lower cost of nickel, as well as improving the thermal stability of the cell, but at the expense of reducing the rate capability of the cell due to reduced lithium diffusion rate.
- *Lithium nickel/manganese/cobalt oxide (NMC):* NMC cathode material is a combination of nickel, manganese, and cobalt oxides. NMC has been used successfully in hybrid and electric vehicles due to its high energy and power density, respectable cycle and calendar life, better safety than pure cobalt cathodes, and good performance at extreme temperatures. The reduced cobalt content reduces the material cost due to the high cost of cobalt. The ratio of the nickel, manganese, and cobalt can also be varied in different formulations of NMC. The most common variant uses equal amounts of all three constituents and is known as 1-1-1 NMC.
- *Nickel cobalt aluminum (NCA):* NCA cathodes are relatively specialized and used only for specialized applications. A small amount of aluminum oxide is used with a predominantly nickel-cobalt blend ($LiNi^{0.8}Co^{0.15}Al^{0.15}O^2$ being the most common formulation). The energy capacity of NCA is high and cycle life characteristics are better than LCO, and NCA is available at less than half the cost of LCO material. The safety margin of NCA is slightly better than LCO but worse than all other common cathode materials. The voltage of NCA cathodes is lower than LCO cathodes.

- *Lithium manganese oxide (LMO)/carbon:* Lithium manganese oxide has the highest cathode voltage of most commonly available cathode materials, which makes the cell voltage for manganese cathodes very high, approaching 4.2V at the fully charged state. Power capability of LMO-based cells is very high due to the extremely low impedance of this material. This occurs because the pathways for lithium insertion and deinsertion are a three-dimensional structure rather than two-dimensional as found with LCO and LNO. Manganese-based cathodes suffer from poor calendar and cycle life due to capacity loss, especially at elevated temperatures. The capacity loss occurs due to manganese dissolution into the electrolyte.

- *Lithium iron phosphate (LFP)/carbon:* Energy density for LFP cells varies between 90 and 140 Wh/kg. These cells would have a nominal voltage of 3.3V and an operating voltage range between 2.5V and 3.75V. LFP has an extremely flat voltage discharge profile over much of the useful SOC range, with the voltage being approximately 3.3V from 20%–70% depending on the exact formulation. The flatness of the curve is due to the formation of a two-phase mixture during discharge rather than a continuous reduction in lithium concentration. $LiFePO_4$ is significantly less conductive than other cathode materials in its pure state and various treatments and additives are needed to achieve satisfactory performance. The lower voltage of LFP means that more series cells are needed to achieve a given system voltage and for a given amp-hour capacity the watt-hour content is correspondingly lower. Lithium iron phosphate is significantly more stable than other cathode materials and offers the highest safety of the common cathode materials. The temperature at which thermal runaway occurs with LFP material is higher than transition metal oxide-based cathodes, and the amount of energy evolved during cathode decomposition is lower. The reduced energy density of LFP also implies that an LFP-based system will be larger and heavier than with other cathode materials.

The above cathode materials are generally combined with a carbon-based anode.

- *Lithium titanate (LTO):* Lithium titanate, unlike most of the other materials discussed here, is an anode material. Lithium titanate cells have a very high charge/discharge rate but a low energy density and specific energy (less than half of the most energy-dense cells, usually less than 90 Wh/kg) and a very low cell voltage (as low as 1.8V at full discharge). Because the volume change of the anode material during charge and discharge is small compared with carbon, less internal stresses result and the cycle life capability of lithium titanate-based batteries is very high but the cost is significantly higher than most other cell types. Many LTO cells are capable of more than 10,000 cycles at 80% depth of discharge (could be four to six times greater than other cells). LTO cells often have allowable recharge rates exceeding 10C, or 10 times the rate that charges the battery in one hour, allowing for recharge times of 10 minutes or less, which is attractive for many applications. The operating temperature range of LTO is wider than most other cells. LTO allows rapid recharging without risk of lithium plating as LTO anodes operate at much higher voltages than required for lithium plating, and also avoid the formation of the SEI (solid-electrolyte interphase), and permit the use of electrolytes with higher conductivities. For these reasons, LTO cells can deliver extremely high power density.

These basic materials are often modified through the use of dopants and additives. Dopants may modify the electrochemistry or they may affect the material morphology and structure. These may serve to reduce the reactivity between the electrode material and electrolyte (improves cycle and calendar life), increase the conductivity of the electrode material (improves power capability), or improve capacity. Electrode coatings have been introduced for improving longevity by preventing electrode dissolution or side reactions and have also been shown to improve safety during abuse conditions such as overcharge or crush and penetration.

As the exchange of ions occurs at the interface of the electrode material and the electrolyte, the particle morphology and microstructure of the electrode has a significant impact on performance as well as the chemical composition of the material. Ions must diffuse to active sites in particles where ion change can occur, and therefore small particles (so that the diffusion dimension is small) exhibiting a large surface area in contact with the electrolyte are preferred. Particle porosity must be sufficient to allow access to the surface by the electrolyte. Particle shape also influences a number of factors including surface to volume ratio and can influence battery performance significantly.

Implications for the battery management system include:

- *Voltage measurement range:* Although many lithium-ion chemistries can be covered with a measurement range of 0 to 5V, the specific safe working range of different chemistries is different and voltages that are safe for one chemistry are not safe for another. Some effort has been put into optimizing the measurement range to place more precision in the useful working range of the cell, and this trend may continue as battery management system developers push for more accuracy. If the maximum measurement capability of the hardware does not cover the full range of all of the cell types discussed above with appropriate accuracy, the battery management system will likely need to be modified for different chemistries.

- *Voltage to SOC relationship:* The shape of the voltage-SOC curve will drive a number of battery management system factors. A flat voltage profile during discharge implies that a more accurate measurement circuit and battery model are needed to achieve accurate SOC calculation. It may also impact the fidelity of the battery model required and the degree of simplification that is permissible while maintaining acceptable SOC performance. The addition of dopants may affect the shape of the SOC/OCV curve.

- *Cell internal dynamics:* The specifics will be covered later, but cell design, chemistry and a number of other factors will greatly influence internal cell dynamics including polarization and hysteresis, implying that a different battery model may be required for different cells using the same or similar chemistry. Cells with extremely long-lasting internal polarization may require hardware which is capable of making measurements during long periods of inactivity for accurate state estimation.

- *Recommended temperature ranges:* The battery management system must be capable of measuring temperatures over the entire operating range of the cell. Temperature accuracy may vary over this range, but it is critical to have accurate measurement around critical transition points in cell capability. Measurement accuracy and range are usually hardware-dependent and may need to be optimized for a different cell chemistry with different temperature characteristics.

- *Self-discharge rate:* Cell balancing systems are sized in relation to the self-discharge rate (and differences in self-discharge rate from cell to cell). Cells with different energy capacity, higher self-discharge rates, and/or more variation in self-discharge (due to manufacturing variability, temperature, or age) will require more balancing capability.

- *Cell degradation characteristics:* A challenge for all battery management system developers is maintaining the performance of battery management system algorithms as the fundamental performance of the cell changes due to cell aging. Presently, most battery management systems have only a limited capability for adapting to the increased cell impedance, decreased capacity, increased self-discharge, and changes in dynamic behavior as the cell ages. The calendar and cycle life of the cell must also be compared with the expected service life of the cells in the

application, A cell with a faster rate of degradation, or used over a longer service life will require more adaption capability in the battery management system to keep up with the changing cell performance.
- *Safety:* More reactive chemistries will require a higher level of care to prevent hazards leading to dangerous conditions. This topic will be discussed in detail in the sections referring to functional safety, but it should be assumed that a risk assessment is performed when determining the required reliability of various functions of the battery management system. This risk assessment will always consider the severity of the reaction of the cell to a potential battery management system failure and therefore a choice in cell chemistry, size, or other design factors will influence the battery management system requirements.

## 2.4 Safety

All lithium-ion batteries, despite recent advances in cell technology and safety, present hazards that arise if the battery is used outside of its safe operating area. Particularly in the area of large-format batteries, demonstrations that cells have passed industry standard abuse tests are not an adequate guarantee that the system offers inherent safety and is robust against all types of abuse conditions. The need to use the batteries in a controlled manner and prevent cell abuse is one of the fundamental reasons that battery management systems exist for lithium-ion battery systems. The battery management system must, to a high degree of certainty, prevent the battery from being abused in a way that can result in an unsafe failure mode, and mitigate the hazards associated with a worst-case eventm such as a vehicle crash or exposure to extreme heat.

A brief list of lithium ion battery abuse conditions is:

- *Overcharge:* Overcharge occurs when a cell is charged to a state of charge greater than 100%. The cell voltage rises very quickly and can exceed the allowable limits of the load device or the monitoring circuit. Overcharge causes a number of irreversible degradation mechanisms inside the cell which can lead to an energetic failure. This is true of a single, severe overcharge event as well as repeated minor overcharging. Unlike other battery types, lithium-ion cells can be overcharged by even very low rates of charge current. Overcharge can lead to thermal runaway, cell swelling, venting, and other serious events. The robustness of various cell designs to overcharge varies significantly and should be well understood when the design of the battery management system is undertaken. For most lithium-ion cells, overcharge begins to occur when the cell voltage exceeds a voltage between 3.75V and 4.2V.

- *Overdischarge:* Overdischarge is the discharge of a cell beyond 100% depth of discharge (DOD) (0% SOC). Cell voltage falls rapidly and can even be reversed if the overdischarge current is high enough. Reverse cell potential can cause failure of management electronics and subsequent malfunctions. Overdischarge can lead to significant internal cell damage including dissolution of the anode foil. Subsequent attempts to recharge a cell that has been deeply and repeatedly overdischarged can lead to safety risks. Overdischarge is a particular challenge since the self-discharge of the battery cells cannot be prevented by the battery management system, even if the battery system is disconnected from its load. Minimum allowable discharge voltages vary from 1.8V to 2.5V per cell.

- *High temperature:* Exposure to high temperature, in addition to increasing the rate of cell degradation, can lead to thermal runaway, in which the activation temperature of various exothermic (heat generating) chemical reactions inside the cells is reached and the cell degrades rapidly with a large release of energy, leading to venting of cell contents, temperature increase, fire, or explosion. High temperatures can result from exposure to high ambient temperatures, exposure to abnormal sources of heat or battery overload (excessive charge and/or discharge power levels) leading to internal heating. The range of acceptable temperatures varies, but most cells begin to experience higher rates of degradation above 45°C–55°C and approach safety limitations between 60°C and 100°C.

- *Low temperature:* Most lithium-ion cells have limited performance, especially charging capabilities at low temperatures. Charging at low temperatures can cause plating of metallic lithium on the anode leading to irreversible capacity loss and the possibility of metallic "dendrite" growth, which can penetrate the separator, causing an internal short circuit. Discharge capability is also limited under low temperature due to increased cell impedance. Many cells recommend inhibiting charging below 0°C and some permit low-rate charging down to –10°C.

- *Overcurrent:* Excessive charge and discharge currents can cause localized overcharge and discharge to occur, leading to the same types of reactions as generalized overcharge and overdischarge. High currents also lead to internal heating, which can lead to over-temperature conditions. The maximum allowable current varies widely for different types of cells, is usually different in charge and discharge, and is a function of both state of charge and temperature. Even if thermal effects are kept in check, excessive charge currents can cause lithium plating due to the anode's limited ability to accept lithium ions at high rates.

- *Internal cell defects:* Inclusions of foreign matter, defects in the cell separator, and other internal faults can cause internal short circuits, which can lead to thermal events. Foreign matter that penetrates the separator can create a short circuit, which causes localized heating and further damage to the separator, resulting in further shorting. These faults were highly publicized due to a series of laptop battery fires in 2006–2008, and robust controls must be in place to address them. The risks associated with these defects can be minimized, but the likelihood can never be reduced to zero and appropriate preventive measures should be incorporated to prevent propagation of thermal runaway between cells or modules. Advanced techniques will be discussed that may be able to detect this type of "soft" short defect before it worsens causing a hazard.
- *Mechanical shock, crush, penetration:* Mechanical damage to cells or systems can cause internal or external short-circuiting leading to venting or leaking of electrolyte and cell contents, thermal runaway, or fire and shock hazards due to electric arcing. Ensuring robust safety against mechanical defects is a complex interdisciplinary task with no guarantee that a severe impact will not create a dangerous condition, but best practices dictate that safety and resistance to mechanical abuse are needed at every level of the system development, including cell materials and design, battery modules, pack design, system integration, and battery management system behavior.
- *Age:* While not specifically abusive, the probability of most of the failure modes associated with lithium-ion batteries increases with age. Lithium-ion batteries have a cycle life span varying from 300-800 cycles to tens of thousands of cycles for cells optimized for high cycle life. Calendar life can vary from a few years to 10 to 15 years for cells intended for the longest service lives.

Of these, the battery management system often bears primary responsibility for preventing the first five events (overcharge, overdischarge, high temperature, low temperature, overcurrent) and has a secondary responsibility in ensuring a prompt response to the other three to maximize the safety of the battery system.

In large-format battery systems, failure to prevent these abuse conditions can lead to thermal events including fire and explosion, release of toxic and flammable substances, and electric arc and shock. As a result, the battery management system for large-format lithium-ion batteries must offer robust and reliable protection against these events. The presence of many cells in large-format systems brings about the possibility of propagation of thermal events from one cell to another. Therefore, overcharge or overdischarge of even a single cell is as a hazardous condition. Even if propagation does not occur, the reduction in overall pack performance due to a single damaged cell indicates that all cells must be carefully managed so that no single cell experiences overcharge or overdischarge.

Designers of battery management systems must understand the susceptibility of the cells being used to particular types of hazards as well as the safety requirements and environment of the application. Less robust cells will require more attention to safety in the design of the battery management system. The safety requirements are highly dependent on the specific application. Many battery-powered systems are capable of causing property damage or injury if there is a battery fire or other thermal event. As battery technology improves, lithium-ion batteries will begin to serve increasingly more in critical applications where the battery must continue to supply energy and power to prevent catastrophic failure of the system (aviation and life support are two examples of critical applications).

## 2.5 Longevity

Loss of performance of battery systems, in terms of both cycle life (degradation of the battery with each charge/discharge cycle performed) and calendar life (degradation of the battery strictly as a function of time since manufacturing), is a concern with all battery technologies. Although lithium-ion batteries exhibit better cycle life and calendar life performance than many other batteries, the size and cost of large-format battery systems generally dictate that the battery management system bears some responsibility for maximizing the longevity of the battery cells.

The principal modes of degradation for lithium-ion battery cells are *capacity fade* and *power fade* (also referred to as *impedance growth*). The usable capacity of the battery generally decreases over time, reducing the amount of energy that the battery can store. The internal impedance of the battery also increases with battery aging, leading to a reduction in the available power from the cell.

Although different battery chemistries, designs, and technologies can have widely varying life-spans, the longevity of batteries is affected by a number of factors, including:

- *Temperature:* A number of chemical reactions which lead to both capacity and power fade are accelerated by temperature. Low-temperature charging can cause lithium plating on the anode, which leads to premature loss of capacity.

- *Operating window:* Operation at high and low states of charge tends to be more damaging to the battery than operation at an intermediate state of charge. As a result, batteries that are deeply discharged and fully charged will usually degrade faster.
- *Charge/discharge rate:* Faster rates of charge and discharge will normally lead to faster battery degradation. In many cases the influence on lifetime of charge and discharge rates may be very different.

In large-format battery systems, the battery management system is usually responsible for controlling all of these factors to some degree. Temperature control can range from actively heating and cooling of the batteries, to limiting performance as temperatures reach extremes, to simple high and low temperature shutdown. The battery management system must accurately determine the state of charge and ensure that the desired operating window is respected to obtain the expected lifespan. The battery management system must also communicate, generally in real time, the maximum permissible charge and discharge rates to the load device. The operating parameters for maximum life span are more restrictive than those for safety, and the battery management system has the responsibility to manage an appropriate balance between safety, longevity, and performance. Additionally, it is often helpful to track characteristic parameters about the battery's life such as the maximum and minimum temperatures experienced, number of charge/discharge cycles, and time in service. This information can be used to determine if the battery cells have been exposed to abnormal conditions.

As the battery ages, the battery management system is expected to monitor the remaining lifetime of the battery, a metric usually expressed as *state of health* (SOH). SOH can consider predictive inputs such as the number of cycles and calendar time in service, as well as real-time measurements made on the battery.

## 2.6 Performance

The battery management system in a modern large-format battery system is expected to go beyond the basics of preventing loss of performance and ensuring safe operation. The expectation exists that the user will be able to experience the full potential of modern lithium-ion batteries. The battery management system is therefore required to provide accurate and dynamic feedback about the battery's capability such that the application can use the battery appropriately.

The relatively high cost of battery energy storage increases the importance of optimizing the performance of the battery system. While, in general, battery systems are sized to ensure that adequate performance is available at end-of-life and off-nominal use conditions, the cost of the system must be kept to minimum. A battery with a battery management system that is capable of using the battery closer to its limits will require less oversizing and therefore offer a lower cost.

In certain applications in which battery systems are used to perform functions that cannot be safely interrupted or stopped, accurate understanding of battery capability may be critical for the safety of the overall battery-powered system. Shutting down a battery powering a life-support application or aircraft motor to prevent overdischarge will have serious consequences.

Critical performance parameters expected from a large-format battery management system include:

- *State of charge:* State of charge is used by the load device to determine available run time and is also linked to a number of other battery characteristics such as impedance and power capability
- *State of health:* A model of the effects of aging on the battery is important for many applications to give an indicator of the remaining useful life of the battery as well as modeling the changes in the battery's behavior as it ages.
- *Power limits:* Most "smart" battery applications communicate with the load device to determine the battery's real-time charge and discharge capability.

Inaccuracies in these parameters will lead to a battery that needs to be oversized to ensure that the performance targets are met. A battery management system that accurately determines the battery condition is capable of pushing the battery further and achieving greater performance from the same cells.

## 2.7 Integration

Many modern battery management systems are expected to do more than simply monitor battery condition and calculate performance data. Integrated systems rely on the battery management system to measure data from additional sensors and inputs, and control actuators and outputs which drive auxiliary functions. Systems with a high level of thermal integration may use the battery management system to monitor a number of temperature sensors throughout the system and control fans, pumps, and heating/cooling devices to maintain battery temperature in a desired range. It is common for the battery management system to control contactors and relays to maintain system safety by disconnecting the battery if it determines it to be necessary to do so.

# 3

# Large-Format Systems

## 3.1 Definition

In the simplest of battery-powered applications, a single battery cell powers a load device. A switch or other control mechanism may exist to interrupt the current between the battery and the device. The battery may be of the primary (nonrechargeable) or secondary (rechargeable) type. The load device is generally unable to receive any information about the battery's condition. The load may not operate correctly or at all if the battery is deeply discharged, at a very low temperature, of the wrong type, or connected in the wrong polarity. Examples of such systems include flashlights, battery-powered radios, many simple consumer electronics, and even the 12-V starting and ignition system in many automobiles. Usually the number of individual battery cells in these types of applications is small.

Large-format systems, in the context of this book, are those with two major differences. The number of battery cells is much higher (systems containing hundreds of cells are not uncommon) and the interactions between the battery and the load device are more complicated that the simple electrical connection that provides energy.

Large-format systems (see Figure 3.1) may be created using small-capacity cells such as the common cylindrical 18650 (18-mm diameter, 65-mm length, used primarily in laptop battery packs, capacities of 1.5–3.4 Ah) and 26650 (26-mm diameter, 65-mm length, often found in power tools, capacities of 3.0–5.0 Ah), or any other types of cells. Little standardization exists for lithium-ion battery form factors other than these small cylindrical sizes. Individual cells usually have capacities from 1.5 Ah to 60 Ah. In a given cell form factor,

Figure 3.1  Typical large-format lithium-ion battery system.

trade-offs exist between energy capability, power capability, and life, and the selection of a particular cell is specific to each application.

Some examples of lithium-ion cells used for large-format systems are shown next.

Battery management systems exist for smaller-format "smart" batteries, such as those found in laptop computers and mobile telephones (which consist of a much smaller number of cells), but there are a number of requirements that make these devices unsuitable for large format systems.

Examples of large-format battery applications include:

- Electric vehicles, including hybrid, plug-in hybrid and battery electric vehicles;
- Grid-tie utility scale energy storage systems;
- Backup power systems.

Compared to the typical laptop battery, which contains six 18650 lithium-ion cells, each cell with an operating voltage of 3.7V nominal and an amp-hour capacity of 2.2–2.4 Ah for a total energy capacity of approximately 50 watt-hours, large-format systems will typically have energy capacities in the range of kilowatt-hours to megawatt-hours, and operating voltages of 250–1,200 VDC.

Some typical battery sizing parameters for different applications are given in Table 3.1.

Most, but not all, large-format systems are high voltage. For a given energy storage and power level, electrical efficiency increases as voltages rise and currents fall. The division between high and low voltage systems is not clear-cut, but there are important differences between the voltages encountered in modern large-format battery management system electronics and those commonly seen in other types of embedded control systems. The components selected and design of electronics for high-voltage applications are different than those used in many low-voltage battery systems as found in most consumer electronics devices. Systems operating at 42 VDC or greater should be treated as high-voltage systems throughout the development of the electronic control system.

The level of safety required for large-format systems is also a critical differentiator. High voltages increase the severity and risk level of electrical shock and arc hazards significantly, and therefore the level of care required to mitigate these risks must increase in proportion. Battery currents typically exceed several hundred amperes, which can create significant hazards if these currents flow through high-impedance connections. Large-format systems may store tens to hundreds of kilowatt-hours of energy with hundreds to thousands of battery cells. In the event of a thermal event, short circuit, or electrical shock, the potential amount of energy released is much larger. The typical electric vehicle stores approximately 500 times more energy than a laptop computer's battery.

## 3.2 Balance of Plant

Large-format battery systems contain more than the battery cells and battery management system. A number of other critical components and functions make up a modern battery system.

Contactors or relays are usually used to disconnect and connect the battery from the load device. As such, there is no voltage present on the terminals of many batteries unless they are connected and actively communicating with

**Table 3.1**
Battery Sizes for Various Large-Format Applications

| Application | System Voltage (VDC) | Battery Capacity (Ah) | Energy Capacity | Peak Power |
|---|---|---|---|---|
| Hybrid electric vehicle (HEV) | 250 | 4.8 | 1.2 kWh | 30 kW |
| Plug-in hybrid electric vehicle (PHEV) | 350 | 42 | 15 kWh | 100 kW |
| Battery electric vehicle (BEV) | 350 | 68 | 24 kWh | 100 kW |
| Grid-tied energy storage | 750 | 1,300 | 1 MWh | 1 MW |

a load. These contactors and relays are often controlled by the battery management system to ensure the battery connects to the load when it is required and disconnects when not needed or in the event of a critical fault. Contactor control is considered critical in many cases, as loss of control of this function prevents disconnection of the battery in the event of a hazard.

The battery management system often requires sensors for current and temperature measurement. In some systems, the battery management system modules may measure current directly.

The battery is contained in an enclosure that is appropriate for its application and environment. The enclosure protects the battery cells and prevents against electrical shock hazards by preventing contact with electrically live parts. An interlock system is often used to disconnect or de-energize systems if the enclosure or connectors are opened.

Overcurrent protection devices, both passive and active, are nearly always incorporated. The battery management system may be expected to monitor and control the state of these devices.

A number of new technologies exist for new ways of detecting additional battery hazards are being developed and integrated into modern battery management systems. These methods include detecting cell venting and internal short circuits. These technologies are still largely in development but will form part of an improving safety strategy for large-format systems. The lower energy capacity and high cost sensitivity of smaller lithium batteries limits the application of these technologies to larger format systems.

Isolation or ground fault detection is often a requirement for modern battery systems. This feature detects conditions where users could be exposed to potentially hazardous voltages through contact with normally nonelectrified parts of the battery system. By contrast, the voltages encountered in typical computer or consumer electronics applications are not hazardous to even direct contact and few if any electrical shock precautions are needed.

Other unique features, such as activating fire suppression systems and controlling power electronics, may be incorporated depending upon the application.

## 3.3 Load Interface

In large-format systems, there is more than simply the high-power electrical connection between the battery and load over which the power is provided to operate the load. A method of communication exists between the battery system and the load device by which information can be exchanged. Typical types of information include:

- Battery state of charge, state of health, and charge and discharge limits;
- Requests to connect or disconnect the battery, or notification that the battery is going to connect or disconnect;
- Values of measured quantities such as cell and pack voltages, temperatures, current, and power;
- Commands to control auxiliary parts of the battery system (cooling fans, pumps, and so forth);
- Status of other devices in the system.

In a high-power grid-storage application used for frequency regulation, the battery can see high rates of charge and discharge (5C–10C) capable of fully charging or discharging the battery in a matter of minutes. The battery current can vary between rapid charge and discharge within seconds. Responsiveness is a key feature of the battery management system for this type of application, capable of calculating the battery's capabilities every 50–100 milliseconds.

## 3.4 Variation and Divergence

In many large-format systems, the assumption should not be made that all battery cells are identical. Some variation in the manufacturing process will lead to variability in capacity, impedance, self-discharge, and other parameters. This variation may increase with age, due to latent effects of manufacturing variation that manifest themselves over time, and also due to the differences in the way that individual cells are operated. Large-format battery systems should be designed to minimize both of these effects

As such, the following assumptions should be made with most large-format applications:

- The capacity of all battery cells is unequal. Although modern battery manufacturers are achieving higher uniformity and quality, no two cells will be exactly identical.
- The capacity of all battery cells is unequal to the rated capacity of the battery cell. Large-format systems generally operate for periods of time that are long enough where the cells will begin to display capacity loss. Importantly, not all cells will display capacity fade at the same rate, depending on the way the cell is aged and manufacturing deviations from cell to cell.
- The self-discharge rate of all battery cells is non-zero. Although lithium-ion batteries exhibit self-discharge rates that are an order of magnitude

lower than many other battery technologies, it is not safe to assume that the rate is zero.
- The self-discharge rate of all battery cells is not equal. It should never be assumed that the cells lose charge at the same rate. As such, even with identical capacities may diverge in state of charge depending on differences in state of charge.

For all the reasons above, it is recommended to assume that the state of charge of a number of "identical" series-connected cells which have all been exposed to the same current profile is not equal.

A large-format battery management system must compensate for these nonideal characteristics to prevent divergence of the state of individual cells. The battery system may be expected to operate for years without service or interruption. Minor differences in cells must not accumulate into large changes in performance over time.

This leads to a number of additional functions that the battery management system must perform. Cell balancing is required to obtain maximum performance from the battery system by equalizing the state of charge between series connected cells. Power limit and state of charge algorithms must consider the facts that not every cell is identical and that these outlying cells will be the first ones to reach the edge of the safe operating area. The limiting cell or cells may change with time and with operating mode (the limiting cell in charge may not be the most limiting cell in discharge). For all of these reasons, it is important that large-format battery management systems are capable of managing a battery pack as collection of individual and different cells. Differences in manufacturing and exposure must be controlled and managed for a long service life. Generally, these differences are small and a challenge for cost-effective battery management system design is to provide an adequate level of performance without excessive cost.

## 3.5 Application Parameters

In the same way that specific cell characteristics will influence battery management system design, there are many aspects of the intended application that will impact the design of the battery management system in fundamental ways.

The "size" of the battery system is the first set of parameters to be understood. The battery pack will consist of a number of series elements strung together, each of which will consist of one or more cells in parallel. The total energy and charge content of the system are a function of the individual cell's energy and charge content (which are defined by its geometry, internal construction, and chemistry), multiplied by the product of the series and parallel cell counts. In many cases, the actual energy used by the application will be restricted significantly from the total energy content of the cell; this is done to increase the cycle and calendar life of the battery system, and/or to increase safety margins against overcharge and overdischarge, or to provide reserve capacity in an emergency. The voltage of the system is a function of the individual cell voltage range, which depends primarily upon the chemistry selection, and the number of series elements. For a given cell capacity and number of series elements, a larger number of parallel cells will increase the charge capacity as well as decrease the overall string resistance, increasing power capability. The cell chemistry, capacity, series/parallel (often abbreviated with s/p notation) arrangement, system voltage range, and required energy and power availability are required for any battery management system development program. If the battery management system is intended to serve a range of possible configurations of the above parameters, the expected design space must be fully defined. A well-designed configurable battery management system can make the configuration process simple and intuitively linked to physical parameters such as number of cells in series and parallel by performing straightforward scaling calculations internally.

The type of environment must be well understood. Battery management system components operate at high voltages and must be maintained clean and dry. If the environment does not meet these requirements, the battery management system will need to provide its own level of protection against intrusion of moisture and dust. Transportation applications will expose battery management system components to mechanical shock, vibration, and g-forces. Applications such as aerospace will have strict constraints on system mass. Most commercial and industrial applications will have requirements for electromagnetic compatibility and interference.

Requirements for serviceability are a factor that may significantly impact design choices. Unlike laptop and mobile phone batteries, which are often only replaced with the entire device (and consequently the management circuit as well), large-format battery systems may be expected to have service lives of 10 to 20 years or more. During this time, due to their high cost, it is reasonable to assume that battery modules may be replaced, and that battery management system components are expected to also offer serviceability and interchangeability. While consumer electronics often do away with service capabilities in favor of lower costs and small packaging, large-format systems in critical applications will need to have replaceable battery management system components. These systems may have requirements for the maximum downtime that is allowable or minimum availability, and therefore the battery management system and associated connections must be easily disconnected and reconnected with minimum effort. Consider the reliability, uptime, and service life of the system for any large-format battery management system design.

High-reliability systems will also drive internal changes as well. Systems that are expected to operate continuously for many years have specific design challenges associated with microprocessors, which are expected to operate constantly without data or program corruption that are not faced by typical battery power systems. Batteries for backup power for critical applications may operate only infrequently during periods of power loss, but the expectation is that the system will never fail to respond, and therefore reliability remains a principal concern.

Special standards apply to a number of different possible applications for large-format batteries and the associated management systems. Battery systems for uninterruptible power supplies, integration of distributed and renewable generation sources, and other types of stationary energy storage may be expected to meet UL1973. This standard was developed for rail applications and is still applicable in that industry as well.

High-voltage automotive batteries often conform to a number of industry-wide standards, as well as stringent requirements from the specific manufacturer.

If isolation detection is performed in automotive applications, it should be done in accordance with the FMVSS 305/SAE J1766 standard.

Battery management systems for aviation have failure modes that can lead to dangerous failure modes that can endanger the airworthiness of aircraft, and therefore they are subject to the DO-178B standard for aerospace software.

Battery management systems intended for automotive applications are electrical/electronic systems and fall under the scope of the ISO 26262 standard for functional safety.

Battery discharge and charge rate can vary from a full charge over many hours to a complete discharge in only a few minutes. Charge and discharge rates have an impact on battery management system requirements. Fast charge and discharge rates lead to higher levels of polarization and hysteresis. This may cause battery models that are accurate at low rates to become inadequate at faster charge and discharge rates. Conversely, low rates increase the error associated with amp-hour integration, which also requires higher model fidelity.

Different applications will require different levels of accuracy for state of charge and state of health estimation. Applications that are expected to use a large portion of the battery's available energy will need more accurate SOC estimates than those with a generous margin at the end of charge and discharge. If estimates of the remaining energy are critical, then the state of charge and capacity estimation must be correspondingly better. This can lead to more advanced models requiring increased processing power, higher measurement accuracy, and faster measurement frequency.

# 4

# System Description

A black-box (input/output only) view of a battery management system is an important step in the development process. It should describe all of the interfaces that the battery management system will have with both the battery cells, additional battery components such as sensors and contactors, and the host application.

The external interface containing the highest volume and rate of information is that between the battery and the load device (which may be a single device that charges and discharges the battery, such as a bidirectional inverter in a grid-tied energy storage system) or multiple devices that combine to perform these functions (referred to as a *load network*; a good example is the powertrain of an electric vehicle complete with motor, inverter, and battery charger, as shown in Figure 4.1. Modular battery management system implementations (see Chapter 5) will also involve data-rich interfaces between individual battery management system components.

It is often helpful to draw separate diagrams for physical and logical interfaces. Each logical interface or signal consists of one piece of information that flows into or out of the battery management system. Examples of signals could include battery pack current, cell voltage, or desired status of a relay.

Physical interfaces are composed of one or more electrical circuits. In the case of communication buses (CAN, RS-232, Ethernet), a single physical circuit may carry multiple logical signals. In the case of a discrete or analog input or output, a physical interface may only carry one signal. Fail-safe or redundant control circuits may use two or more physical circuits for the same logical function.

The electrical parameters of each physical interface should be clearly defined. For example, an analog input will have a voltage range, input impedance and maximum applied voltage. A digital output will have a voltage range for

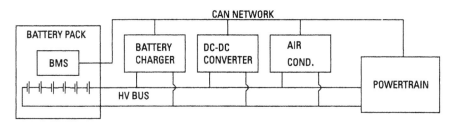

Figure 4.1 Battery management system network in a typical electric vehicle.

both the logic 0 and logic 1 states, a maximum current which can be supplied and a transition time from on to off.

The concept of information hiding is a best practice in software engineering and is a useful tool in systems engineering as well. Each subsystem should communicate to other subsystems only the information that is required for their operation. In the context of battery management systems, it is often assumed that the load device and other components on the network require all of the information used by the battery management system to perform its internal calculations, such as the full set of cell voltages and temperatures. However, an effective system architecture will provide only the information needed to each device.

In the development of large-format battery management systems, it is often important to identify the voltage reference for electrical signals, since there may be multiple galvanic isolation barriers throughout the system. For example, a particular voltage could be referenced to *earth ground*, which would be the potential of the chassis or casing of the battery system, or the high-voltage stack, where the negative terminal of the lowest-voltage cell is usually considered the reference potential, and which is isolated from the chassis in many cases. Other floating voltages not referenced to these two potentials are also possible. For earth ground reference signals, a distinction is often made among analog, digital, and power grounds, which are referenced to an equal potential but kept separate to prevent sensitive analog signals from being disturbed by fast digital switching transients or switching of high-power devices.

## 4.1 Typical Inputs

Inputs can take the form of measurements (physical quantities that exist in the system that the battery management system is measuring) or commands (usually logical quantities that are represented by a physical signal that the battery management system must interpret as an instruction). Some possible signal types are shown in Table 4.1.

**Table 4.1**
Ground and Signal Types

| Ground Type | Typical Signals |
|---|---|
| Analog | Current measurement (isolated), analog inputs for fan speeds, temperatures (isolated) |
| Digital | Serial communications (RS-232, CAN), pulse with modulation (PWM), and timer inputs |
| Power | Contactor and relay control signals |

The battery management system should measure the three basic parameters of battery cells in order to determine the battery state: cell voltage, string current, and cell temperature.

In most cases, the battery cells are connected directly to the battery management system and the cell voltages are measured directly without the aid of any intervening sensors. Due to the high accuracy required (from 1-10 mV total error is a common objective) and need for synchronous measurements, external sensor devices (commonly used for measuring other quantities that need to be converted through some transfer function to a voltage signal) are not used. Depending on the architecture used, the number of inputs for a battery pack with $n$ series cells will range from $n + 1$ (each cell requires a positive and negative connection) to a higher value if the voltage measurements are divided into modules, or if the battery management system is designed for multiple battery pack sizes with potentially a larger number of cells. These inputs are individually relatively low in voltage, but they comprise a large high-voltage stack, and often operate at hazardous voltages relative to other voltages in the battery pack and ground potential. The cell voltage measurements are used to avoid overvoltage and undervoltage conditions, calculate SOC and SOH, calculate and enforce current and power limits, and detect battery failures. The logical and physical interfaces for the voltage signals are the cell voltages themselves. The battery management system design and implementation should aim to create a unity-gain transfer function such that the voltage measured at the battery management system is exactly that which is present at the terminals of the battery cells. When the cell voltage measurement circuit shares components with the cell-balancing circuitry, meeting this goal becomes more challenging.

Measurement of the complete battery series string voltage, as well as module or substring voltages, is also commonplace. As will be discussed later, this additional layer of measurement provides a significant benefit in terms of failure detections.

Battery current is measured by the battery management system through one of the methods discussed in Chapter 6. The current is bidirectional (charge and discharge will occur during battery operation). While there may be many devices on the high-voltage bus, some of which may only be capable of one function (charging or discharging), the measurement of the battery current should be taken at a single point that encompasses the currents flowing to all load devices. Often an external sensor is used to convert the battery current (which is usually very large) to a voltage signal representing the current level. The physical interface in some cases could involve passing the entire battery stack current through the battery management system, but this approach is more common in smaller-format batteries with lower currents. While the logical interface is a signal that represents the current, the physical interface is a voltage signal. This signal may be referenced to the high-voltage stack or to earth ground, depending on the measurement technique used. The signal is expected to change as quickly as the battery current.

Temperatures are also typically measured with external sensors that are physically distributed throughout the battery system at locations where temperature measurements are needed. These sensors typically change in resistance or voltage in response to the temperature measured. The range of the signal must be appropriate in both the temperatures expected and the voltages produced by the sensors at these temperatures. Temperature measurements may be earth referenced, high-voltage stack referenced, or floating.

The load device will usually need to communicate a minimum of information including when the battery should connect or disconnect from the load. This can take the form of a discrete signal or a communications message which requests the battery to enter an active mode. A handshake mechanism usually is implemented where a command is issued by the load network, and the battery attempts to connect to the high-voltage bus. If the battery condition is nominal and the connection sequence proceeds normally, a response will be provided indicating that the battery is ready to operate; otherwise, an error will be generated and a negative response will be returned. See Figure 4.2 for an example of a typical handshake sequence.

## 4.2 Typical Outputs

Battery management systems are typically responsible for computing battery state parameters, which must be determined through complex functions of the input quantities.

Many battery management systems have been designed to output large numbers of cell voltages and temperatures to a host or load device. In many cases, the external device is not capable of processing this information in a meaningful way. Arguments can be made that another device could be used for redundant monitoring and provide additional safety, but it is a poor system design to make a component outside of the battery system responsible even in

## System Description

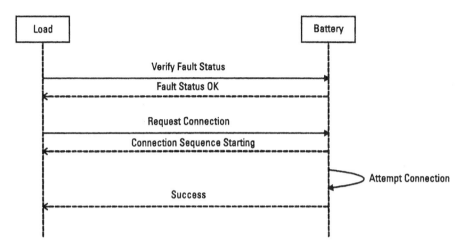

**Figure 4.2** Sequence diagram for battery connection.

part for battery safety; the battery management system should be capable of providing all necessary protection. This type of information may be needed on an occasional basis for system diagnostics, but in general, this type of information does not need to be exchanged with external devices on a continuous basis. Output information should be limited to that which is used in a useful way by external systems.

State of charge is the ratio of the amount of electrical charge currently available for discharge from the battery to the maximum amount of electrical charge that can be stored. It is expressed as a fraction between 0 and 1. In many systems this is converted to a percentage, and may be computed with a resolution of between 1% and 0.1%; much less than this is usually impractical due to the state of the art available for SOC calculation.

State of health (SOH) is a quantity that describes how much of the battery's original capabilities have been maintained. SOH is affected by parameters such as impedance and capacity, which degrade over time. Ideally a new battery has an SOH of 100% and a battery which needs to be replaced because it can no longer meet the application's requirements has an SOH of 0%. SOH metrics are highly dependent upon application, chemistry, and battery requirements, but, in general, refer to degradation in capacity, increase in impedance, and increase in self-discharge rate. A 1% resolution is acceptable in most applications for SOH resolution.

The limits of the battery pack are usually communicated to the load device. The battery has a maximum allowable current and power in charge and discharge that will ensure that the battery remains in a safe condition. These limits change dynamically with the state of charge and temperature over the course of a single cycle, and also changes as the battery ages and becomes reduced in performance. These may be communicated as absolute values (in watts or amps) or as percentages of rated maximum.

Faults and error codes are usually communicated to external devices, including not only the normal load network but also diagnostics equipment used to troubleshoot and identify battery system problems. These fault codes should identify not only the nature of the problem, but also the smallest level of component, cell, or subsystem that is determined to be affected. These types of diagnostics messages may not be part of the normal output set; they may only be issued upon request by a diagnostics device.

## 4.3 Typical Functions

The battery management system is responsible for maintaining cell balance in a large-format battery system, ensuring that all cells are available to deliver and accept a similar amount of charge throughout the life of the battery system, encompassing differences in cell impedance, capacity, and self-discharge. The cell balancing outputs are in the form of currents that selectively charge or discharge individual cells in the high voltage stack. These currents are calculated from cell voltage measurements and are influenced by many factors such as the mode of operation of the battery system. Balancing currents are small in comparison with the full-scale current of the entire battery pack.

Battery management systems are often required to control contactors and relays to connect and disconnect the battery to the load device as needed, including emergency disconnection in the case of an unsafe battery condition, soft-start or precharge functionality for capacitive loads, and discharge of high-voltage buses for safety purposes. These signals could be control signals (low-power logic only) or power signals that are of a large enough current and voltage to control these devices directly. The latter approach increases the battery management system size and power requirements but has a number of benefits in terms of safety.

Advanced algorithms are implemented to estimate the battery's state of charge, SOH, and limits that will be reviewed in later chapters. The time-sequenced measurements of current, voltage, and temperature are processed by complex models to obtain these outputs. These quantities are communicated on a continuous basis to the load network to ensure that the battery system is not used in an unacceptable way. Serial communications, such as CAN, are often used to deliver this information, but analog or pulse width modulation (PWM) signals could be used alternatively or to complement the digital communications link.

The battery management system is often integrated with other functions designed to improve or preserve the performance of the battery system. Thermal management is a good example; battery temperature and other measurements can be used to command drive signals for pumps, fans, chillers, or heaters used to maintain battery temperature at an optimal level for improving battery life. These signals may be digital (on-off control only) or PWM or analog to adjust outputs or speeds. Communications signals may also be used to pass information to external control modules, which will, in turn, create drive signals.

## 4.4 Summary

The inputs, outputs, and processing to be performed define the interface between the battery management system and the outside world, and the features to be implemented by the battery management system designer. At this stage, the description of the internal operation is relatively abstract with little distinction made between hardware and software elements or description of the method of implementation.

A good battery management system will generate accurate outputs from the inputs over the full range of expected operating conditions and environments. Factors about the external environment such as temperature, age, and part-to-part variation, which may influence the accuracy of the outputs, are known as *noise factors*. Each function will have a number of noise factors that affect its performance; for example, battery current measurement may depend (undesirably) on the ambient temperature.

To counteract the effects of noise factors, *control factors* are chosen by the designer to improve the robustness of the function. In the above case, the type of sensor used to measure battery current will influence the degree to which ambient temperature affects the current measurement.

A concise yet complete description similar to Figure 4.3 of all of the interfaces is necessary to begin the design process. Do not forget the mechanical (mounting and sealing interfaces) as well as thermal interfaces to completely specify the battery management system context.

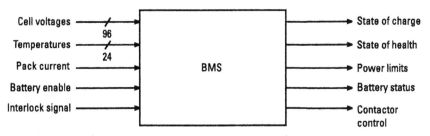

**Figure 4.3** Sample battery management system interface diagram.

# 5

# Architectures

When the system's external behavior and interface have been defined, decisions about the implementation of the system can begin. It is a foregone conclusion that the battery management system will be realized through the use of general-purpose microcontrollers or microprocessors implemented with the supporting measurement, power and control circuitry on one or more printed circuit board assemblies (PCBs), with the control, calculation and analysis functions residing in the microcontroller software. It may be possible to implement the simplest systems using a selection of definite-purpose integrated circuits (ICs), but this architecture is more commonplace with smaller systems operating at lower voltages intended for consumer electronics applications.

A common decision to be made for battery management system development (as with many embedded systems) is the degree of modularity desired.

## 5.1 Monolithic

The simplest solution is to place all of the functionality into one module. Monolithic systems reduce the need for the design, definition, and cost of interfaces between modules. Figure 5.1 shows a monolithic BMS, complete with a large format battery string.

Scalability is limited with a monolithic system. The number of cells which can be monitored is bounded by the number of cell-monitoring circuits installed. In many cases it may not be possible to monitor any arbitrary lower number of cells. Additionally, there are no cost savings realized for smaller batteries because the number of battery management system components cannot be reduced.

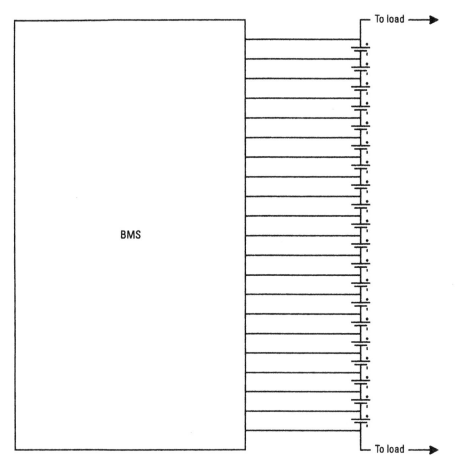

Figure 5.1  Monolithic battery management system.

Monolithic systems require the single controller to support the entire pack voltage and all the cell measurement connections. The creepage and clearance distances must be larger due to higher voltages. Connector and component ratings must also be appropriate and the number of possible component choices for a given application may be restricted. Although a good monolithic design will attempt to reduce the potential difference between adjacent component choices and establish adequate isolation barriers as needed, under fault conditions, the potential for much higher voltages and fault energy exists.

Monolithic architectures are logical when the same battery management system will be used on a large number of systems with only small differences. In the case of very large volumes, a battery management system offering only the required features will provide the lowest cost and complexity at the expense of flexibility and scalability.

## 5.2 Distributed

Distributed battery management system architectures achieve a high degree of modularity. Many systems are built in a master and slave architecture. The most common designs incorporate a single central control module (sometimes called a battery control module (BCM), battery pack control module (BPCM), battery electronic control module (BECM), or battery management unit (BMU), which is responsible for most of the computational requirements, and a number of similar or identical slave modules that are connected to battery cells/modules (sometimes called cell supervisory circuits, or CSCs). The slave modules are responsible for measurement of cell voltages and temperatures and reporting this information to the master device, and also for executing cell balancing under direction of the master.

In a distributed architecture there is usually a strict correspondence between the number of battery modules and slave devices. Systems with a high degree of integration may have the sensing slave circuit incorporated directly into the battery module.

The master and slave devices must communicate, usually using a communications protocol. This protocol could be proprietary or a commonly available protocol such as CAN, RS-232, or Ethernet.

Due to the larger number of communications circuits as well as support circuits such as microprocessors, power supplies, and isolation, distributed systems often have the highest cost. These extra circuits also increase the weight, size, and parasitic power consumption of the battery management system, in comparison with a similar implementation of a monolithic architecture. In the applications in which they are used, this is offset by the advantages discussed above.

## 5.3 Semi-Distributed

A semi-distributed architecture uses a smaller number of sensing circuits which are not integrated with the battery modules. This allows the system to be more easily scaled if the size or form factor of the battery modules is changed. The system is more expandable than a monolithic system and may be expanded with different sizes or types of battery modules without redesigning the battery management system hardware, unlike distributed systems, which are tightly integrated with each module. Figure5.3 shows a semidistributed BMS architecture.

It is possible to further distribute the functionality. A possible architecture involves measurement slave devices, which operate at a potential referenced to the cell stack with an isolated communications interface to a low-voltage only master device operating relative to earth ground potential. These two devices

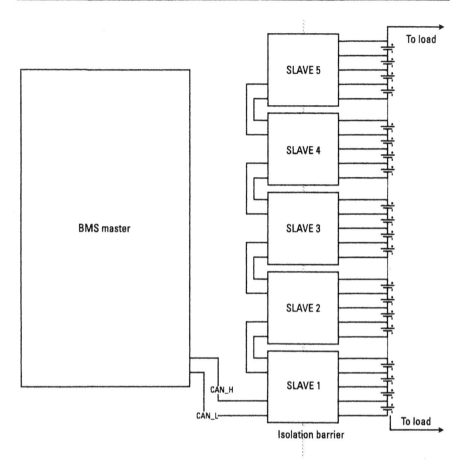

**Figure 5.2** Distributed battery management system architecture.

operate together with a high-voltage measurement and control device that is the only piece of the electronic system exposed to full battery voltage. This module would possibly contain high voltage measurement, contactors or relays, fuses and current sensors, or shunts. The high-voltage module also contains an isolated communications interface to the master module. This minimizes the size and number of devices required to handle high voltages but increases the number of isolation barriers that are required, the number of modules, and the amount of information transported on communications buses. This is a good architecture for extremely high-voltage systems (1,000V or more) in which critical attention is required to the high-voltage elements and the feature size, connectors and insulation ratings required at these voltages may make it difficult to integrate the high-voltage section with the low-voltage control electronics.

*Architectures* 63

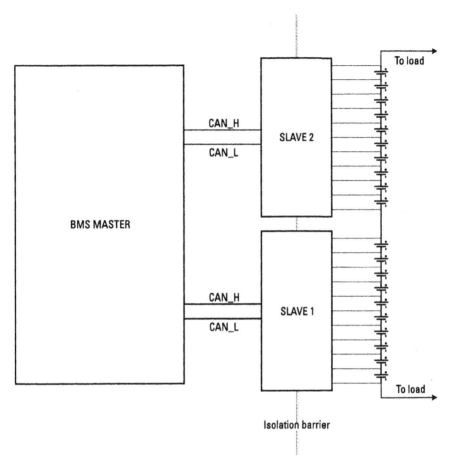

**Figure 5.3** Semi-distributed battery management system architecture.

## 5.4 Connection Methods

Connecting the battery management system electronics to the battery cells is a seemingly mundane task but one that must be accomplished with a very high degree of reliability. This connection performs two functions, carrying the cell balancing current, as well as providing the voltage signal to the measuring circuit.

Wire harnesses are the most obvious method that can be used for this purpose. The wire must be sized appropriately to carry the balancing current without excessive temperature rise. The wire selected must have a temperature rating exceeding the maximum service temperature plus the expected temperature rise during balancing. The failure mode of short-circuit currents flowing in the sense harness is an important consideration. Harnesses may experience shorting due to chafing within the harness bundle, in which case the voltage of a short (which determines the maximum short current) is limited to the maximum voltage of the cells connected by that particular harness or bundle, or a short could develop in the battery management system itself, meaning the short could occur across the full battery pack voltage. The normal expectation is that the harness will act as a fusible link in this situation, but the wire size should be minimized to reduce the total short energy, and appropriate testing is required to ensure that the failure is of a benign nature. The choice of connectors, wire insulation material, routing, and bundling also are important factors to consider when designing a harness to protect the system from a short in this way.

Wire harnesses should be constructed using conductors rated for the maximum expected potential difference between both the battery pack terminals when fully charged, and between the high voltage system and earth ground.

Misrouted harnesses can expose measurement circuits to voltages that exceed the circuit's design limits. In production systems, a comprehensive test—performed before final connection of the sensing harness to the BMS—to ensure that the battery management system is not connected improperly is a best practice to prevent damage to the measurement circuit, or possibly worse.

In the common situation in which temperature sense connections are routed alongside those for voltage sensing, a possible failure situation is unwanted connections between voltage and temperature sensing leads. If the temperature sense circuits are referenced to the cell stack, this can limit the voltage seen across a short. If the temperature sensors are referenced to earth ground, cut-through of voltage and temperature leads can cause not only a high-voltage short circuit but also an isolation fault. If a floating reference is used, the safety risks of such a failure are minimized, but the isolation barrier between the floating reference and the cell stack must be adequate.

Connectors for cell sensing need to be rated for adequate current (equal to the cell balancing current) at rated temperature and most importantly for adequate voltage. When assigning the pinout to the connectors, a layout that minimizes the voltages between adjacent pins should be selected. Do not forget that the connector pins must also respect creepage and clearance distances. Unsealed connectors will require larger creepage and clearance distances if there is a possibility of humidity or dust contamination. A monolithic architecture will have the most connections and highest voltage on the same connector.

Balancing currents are conducted by the sense interconnect when balancing is taking place. In many systems, balancing currents are extremely low and flow only intermittently. Contacts that do not carry continuous current are subject to contact oxidation at terminals (including where the connection is made to the battery modules themselves) leading to high contact resistance and potential measurement issues, especially during balancing. It is important that the balancing current is high enough, and flows frequently enough to remove this oxide growth (this is often called a "wetting" current) to prevent this problem from occurring. Methods of preventing this include:

- Using gold-plated contacts if the cost penalty is acceptable;
- Using sealed connectors to prevent humidity; moisture increases the rate of insulating oxide growth;
- Minimizing the number of connections in the measurement signal path;
- Ensuring that a balancing current flows on a periodic basis for each connection in the battery pack, even if that particular cell does not require any balancing;
- Setting balancing current high enough to provide adequate wetting current (this is not always desirable as it adds cost through larger balancing switches, resistors, and other components);
- Not using mechanical or electromechanical switches in the measurement signal path or to control cell balancing.

Other, more advanced types of integration are possible, including connecting measurement PCBs directly to battery tabs or busbars. These methods often seem elegant and propose significant benefits from the reduction of cable complexity. In many applications, however, this claim should be carefully examined as wire harnesses are simple to build and test, low in cost, reliable if properly installed, and easy to service. Busbars and tabs may also conduct damaging mechanical loads and vibration from battery modules to circuit boards unlike flexible wire harnesses.

Locating the battery management system close to the battery cells usually implies that the location will be both moderate in temperature due to the limited tolerance for extreme heat and cold of lithium-ion batteries as well the need for as a clean and dry environment. Remotely locating the battery management system may lead to more extensive requirements for temperature tolerance and environmental sealing.

## 5.5 Additional Scalability

The number and type of measurement and control circuits for additional measurements (other than current, voltage and temperature) also affect the system's flexibility and scalability. For a system intended to serve a wide variety of applications, it is common to include a number of multiple-purpose inputs and outputs, as well as providing for the largest possible balance-of-plant integration expected (number of contactors and sensors). This adds cost to the electronics but reduces the number of configurations which must be developed, validated, and supported.

Questions to answer in choosing the system architecture include:

- How many cells does the battery management system need to service? How much is this number likely to vary in the range of applications? What is the minimum number and maximum number, and what is the expected step size for adding additional cells? Not every architecture can handle arbitrary numbers of cells.
- What are the types of chemistry of the cells that could be expected in the battery system? What range of state of charge will the cells be operated in and what voltage range?
- What is the maximum and minimum expected battery pack voltage?

## 5.6 Battery Pack Architectures

The battery cells themselves can be divided a number of different ways. The most common solution is a single-string architecture in which the number of battery cells required to achieve the desired amp-hour capacity are placed in parallel to create a "cell group" or series element. This effectively behaves as one large cell with a capacity equal to the capacity of the individual cells multiplied by the number of cells placed in parallel. The required number of series elements is then installed in series to achieve the required system voltage.

It is also possible to place strings of smaller series elements (down to a single cell) in parallel at the pack level. A multiple parallel string system requires more advanced decision making about whether all strings are operating correctly, and adds complexity related to balancing state of charge between strings, but offers redundancy if each string is capable of providing enough power and energy to the load to operate, possibly at reduced capability.

The multiple parallel string topology also mitigates a significant failure mode if a short circuit develops internal to a cell. By reducing the number of cells in parallel, the equivalent source impedance of the cell group increases, but the multiple parallel strings allows the battery to provide the same overall power level. If a short circuit occurs, the current fed into the cell by the other parallel cells in the string will be reduced significantly due to the much higher source impedance. Contractors can be used to isolate strings from each other. This can allow for increased safety with very low impedance cells.

Decisions about battery pack architecture affect the battery management system design in the following ways:

- A larger number of series elements increases the number of cell voltages per system, which must be monitored. There is a component of the battery management system cost that scales approximately linearly with the number of cell monitoring channels, so minimizing the number of cells to be measured will reduce the cost of the battery management system. There are usually some "discretization" effects; to reduce the battery management system cost, the number of cells must reduce by some minimum number so that an extra IC, module, or circuit can be eliminated.
- Multiple parallel strings will require a separate series of monitoring channels for each parallel string and therefore will increase the battery management system cost significantly.
- Higher capacity battery cells require proportionally larger balancing current to achieve the same compensation capability.
- A larger number of series elements, or a higher-voltage chemistry results in a higher total system voltage, which will require larger creepage and clearance distances, heavier insulation, and isolation components with higher voltage ratings. Some of these effects are continuous or semicontinuous and others are discrete, most notably the isolation voltage ratings available for many types of electronic components.
- Multiple parallel strings also increase the number of contactors that must be controlled and the number of high-voltage measurements that need to be made.
- Defining the overall limits, state of charge, and state of health, in a multiple parallel string system is more complex than with a single parallel string.

## 5.7 Power Supply

The power supply for the battery management system itself can come from a few possible sources. The battery management system could be powered directly from the cells in the battery stack, or power could be completely provided by an external control voltage supply. Alternatively the high voltage bus, either on the battery side or the load side, could be used as a potential power supply. Combinations of the above implementations are also possible. There are important trade-offs in the selection of a power supply topology.

Powering the battery management system from the battery cells ensures that the battery management system is never connected to a cell stack without having a source of power to perform measurement and control functions.

If this topology is chosen, it is important that the power consumption be minimized, in both active and passive modes (not to be confused with the measurement current, which also should be minimized). Additionally, if the power consumed by individual modules or circuits varies from module to module, creating an unequal power load on all of the cells, this topology will cause increased imbalance to occur. It is important not only that the power consumption be low but also that the drain from all the cells is equal. Consistency in power consumption from unit to unit is not always a goal of circuit design and this should be considered with battery management system implementation. Because power consumption of electronic devices often depends on temperature, temperature imbalance throughout the system can aggravate these effects. The cell voltages themselves may have an impact on the power consumption as well. Care must be taken, especially with low-capacity systems, to ensure that a scenario does not exist whereby the use of the battery cells for battery management system power overcomes the balancing capability of the system.

Powering the battery management system from the cells will increase the apparent self-discharge of the battery cells due to the constant power draw for the battery management system circuits. It is the challenge of the battery management system designer to ensure that this increase is not unacceptably high.

Using the high-voltage bus or the cell stack may increase overall system efficiency if low-voltage control power is provided ultimately from the battery through a power conversion device. Automotive applications using a dc-dc converter to maintain a traditional 12-V system, which is used to power the battery management system, are a good example of this. In an electric vehicle, the only source of energy is ultimately from the high-voltage battery system. Any energy that is drawn from the 12-V system must pass through a power conversion device, possibly lead-acid storage battery and other components and losses occur when energy is converted in this manner.

## 5.8 Control Power

There is nearly always a requirement for low-voltage dc control power for a battery management system that controls auxiliary devices, contactors, and powered sensors. The 12-V automotive system is a good example, but other voltage ranges may be possible.

In the case of vehicle systems, it is common that this control power is provided by a starting, lighting and ignition (SLI) battery that may be shared with other vehicle systems and, in the case of a hybrid vehicle, used for high-current loads such as engine cranking. This implies that a very wide range of voltages may exist. For example, devices operating from automotive 12-V systems often need to demonstrate functionality as low as 6V (during cold engine cranking) up to 18V (during charging with an unregulated alternator).

It may be desirable to produce the control power by power conversion from the high-voltage stack.

Regardless of its source (internal or external), real-time measurement of this supply voltage is normally required within the BMS for a number of purposes. A variety of functions may need to be inhibited if the supply voltage falls outside normal ranges.

If the control power fails during battery management system operation there is a risk of undesirable events occurring. Contactors may open uncontrollably under load or chatter if the control power becomes unstable. As battery management often requires that data integrity for values such as state of charge be maintained from one operational period to the next, if a significant amount of data is lost, or worse, corrupted during an uncontrolled power-down, errors may be present during the next operational cycle of the battery. Therefore, it is often useful to have on-board backup power for at least the microprocessor and supporting circuitry such that a loss of power leads to a controlled shutdown with an orderly save of data to nonvolatile memory.

## 5.9 Computing Architecture

As the required hardware can be distributed across different modules or concentrated in a single device, the software and associated processing power needed for the battery management system functions can also be arranged differently.

In a monolithic hardware architecture using only a single processor, there is no room for decision-making; the single processor is required to implement all software functions in a single software application.

In a distributed or semi-distributed master-slave architecture, each slave device will typically (but not necessarily) have a microprocessor responsible for, at a minimum, voltage and temperature measurement as well as cell balancing. It may be advantageous to place additional functionality in these microcontrollers to provide additional computing resources, but there are some limitations to the gains that can be realized from this. Slave modules may not always have access to all system inputs, including overall string current, which would be required to perform detailed battery model calculations to generate state of charge estimates, for example.

A multitier architecture is recommended for battery management system implementations (as well as for all embedded control systems). Software functions can be divided into low-level device drivers and hardware interface routines, middle layers providing implementations of communications protocols and interpretations of physical measurements, upper layers that are responsible for high-level battery computations such as state of charge and power limit calculations, and ultimately a top-level application layer that is responsible for decision-making based on information provided by lower levels. Strict use of abstraction layers and a multitier approach maximizes reusability of code modules. For example, an application that decides to connect or disconnect the battery based upon its state of charge does not need information about how the SOC is being calculated, and in fact it may be advantageous to use different methods of SOC calculation in different applications. Consequently, there is no need for the SOC calculation algorithm to understand the details of how its inputs (temperature, voltage, current) are processed. If this type of layered architecture is maintained, any of these layers can be modified with limited consequences to adjacent layers.

Most software architectures will implement a multitasking environment. This could be anything from a straightforward "round-robin" task scheduler to a full preemptive multitasking operating system. In safety critical systems such as battery management systems, it is necessary to ensure that tasks that are responsible for safety functions such as voltage measurement and associated overcharge and overdischarge prevention, temperature and current measurement, limit calculation and contactor actuation are performed in a timely fashion to ensure prompt responses to hazards. In a multitasking environment where it is possible that tasks are interrupted to perform others, it is vital that safety critical tasks are not interrupted, skipped, or performed late.

Consider the use of active profiling, especially during the development phases, to ensure that task overrun does not occur leading to late or missed tasks.

# 6

# Measurement

## 6.1 Cell Voltage Measurement

With the understanding of the importance of preventing overcharge and overdischarge, as well as the need for accurate information about the state of the battery, accurate voltage measurement is a cornerstone of battery management system operation.

A number of battery voltages could be measured within a large-format system, including individual cells, groups, or modules of cells, all the way up to the entire series-connected string.

It has already been made clear that the assumption that a number of series-connected cells in the same battery system are not necessarily of the same capacity or at the same state of charge. As such, most implementations require that at least one voltage measurement of each series element is made. This is in direct contrast to a number of other battery technologies; for instance, many lead-acid 12-V batteries consist of six series-connected cells and do not even have external terminals for measuring the individual cell voltages—it is assumed that the state of the battery can be determined well enough by observing the total battery voltage.

The cell measurement circuit should present a high dc impedance to the battery cells to minimize parasitic power consumption. Whether a given measurement impedance is high enough depends on the capacity of the battery cells being measured; very small cells will need much higher impedances to ensure that the battery management system does not undesirably increase the apparent self-discharge. In addition to keeping the power consumption low, it is also important that the differences in power consumption from the individual cells do not differ greatly, as this will contribute to increased cell imbalance and reduced battery performance. This is particularly a concern where multiple slave modules are connected to different cells and the modules have varying levels of power consumption. The measurement circuit's impedance should be characterized in both the active mode when measurements are actively being made and the passive mode where no measurements are taking place. Reduction of the passive mode current, also called standby current, quiescent current, or parasitic current, ensures that the battery cells will not be drained by the battery management system when the system is powered down and the battery will have a long standby time without risk of cells experiencing overdischarge due to battery management system current draw. Active mode currents are expected to be higher. Depending on the overall duty cycle of the battery system and battery management system, the capacity of the battery, the expected storage/standby time, and the minimum state of charge at which the battery may be operated, the requirements for the active and passive measurement current can be established. The measurement current flowing through interconnects and various devices will create both a voltage drop between the terminal voltage and the measurement circuit, and also potentially cause deviations due to cell overpotentials from true open-circuit voltage (this has been seen with even very low currents) that can interfere with other battery management system functions. Finally, as one of the battery system's primary functions is to store energy, consumption of energy by the measurement circuit negatively affects the efficiency of the battery system (highly important for renewable energy applications where efficiency is paramount). Reducing the energy consumed by the measuring circuit is a common goal of nearly all battery management system architectures.

It is not uncommon to perform multiple measurements of individual cell voltages. These multiple measurement schemes may be symmetrical (multiple measurements of equal precision and accuracy) or asymmetrical (secondary measurement with lower accuracy than the primary measurement). In the most critical of systems, more than two measurements may be needed to provide the battery management system with the relative plausibility of the measurements; it is a simple task to determine that two unequal measurements of the same quantity indicate a malfunction, but the classic question of which piece of data to use and which to discard arises, requiring additional information. In applications in which the hazards associated with a measurement error can be prevented simply by shutting down the battery system and in which the reliability of the measurements is high enough that the overall battery system reliability can meet its requirements, an asymmetric dual measurement may be adopted.

Additional measurements add complexity and cost. The scheme selected for voltage measurement should be subjected to a fault-tree analysis to obtain an understanding of the risk and safety level that must be achieved with the design.

Individual cell voltage measurement range should cover the range of cell voltages expected under normal conditions. As well, in almost all cases it is desirable to have the battery management system capable of handling cell voltages outside normal operating ranges to ensure that in the event of an overcharged or overdischarged cell, the battery management system will be undamaged and able to respond to the event and prevent further abuse. For many in large-format systems in which multiple cells are placed in parallel, it is common to the BMS will perform a single measurement of the voltage of all parallel cells in a cell group. An important failure mode to consider is cell interconnect failure, which creates a subgroup that is connected to the measurement circuit but disconnected from the series string. In this case, the cell voltage that is measured will be constant and not representative of the actual cell voltages in the series string. Overcharge and overdischarge therefore become possible. Detection strategies for this include checking for zero cell impedance or an infinite capacity (cell voltage is invariant with pack current and state of charge).

Other types of battery voltage measurements are often made. The most common is the complete battery stack voltage. Many systems also measure the voltage of individual modules or groups of series elements. Comparisons between these measurements can be used to detect a number of possible measurement faults. Simply put, the sum of individual series elements should always add up to the bulk measurement of the string or substring. This may be used to detect calibration errors in which all cell measurements (or possibly all cell measurements from a particular measurement module) contain a systematic error. It may be used to diagnose wiring faults with the high-voltage measurement circuits. To use these measurements to prevent overcharge and overdischarge hazards resulting from primary measurement errors requires a careful analysis of the accuracy and precision associated with the different measurement circuits.

Selecting the desired level of cell measurement accuracy is another important decision. In systems in which cell voltage is used to compute state of charge, the relationship between errors in voltage measurement and errors in state of charge can be computed.

Assume that a measurement error $\Delta V$ exists such that for a measured voltage $V_{measured}$, the true cell voltage lies between $V - \Delta V$ and $V + \Delta V$.

$$V_{measured} - \Delta V < V_{true} < V_{measured} + \Delta V$$

A function $SOC(V)$ exists that gives the SOC for a given voltage. The range of possible SOCs associated with this measurement is then:

$$\left(SOC\left(V_{measured} - \Delta V\right), SOC\left(V_{measured} + \Delta V\right)\right)$$

A simple linearization can be performed by computing $dSOC/dV$, the slope of this function at the point of interest. The range of SOCs can then be expressed as:

$$\left(SOC(V_{measured}) - \Delta V \frac{dSOC}{dV}, SOC(V_{measured}) + \Delta V \frac{dSOC}{dV}\right)$$

This shows us that the maximum error in state of charge is proportional to both the measurement error $\Delta V$ as well as the slope of $SOC(V)$. A graphical representation is shown in Figure 6.1.

$dSOC/dV$ varies significantly between battery chemistries and also as a function of the state of charge itself. As such, fixed measurement error may lead to a variable error in SOC over the full range of SOC.

Some sample data for $dSOC/dV$ is given next for different types of lithium-ion cells and at different values of state of charge. The wide difference between cell types in rates of change can be seen in Figure 6.2.

For simple prevention of overcharge and overdischarge and enforcing reactive power limits when cell voltages reach operating limits, a lower level of measurement accuracy could be used. Values for secondary measurement error between 25 and 100 mV are not unreasonable. If the primary measurement has

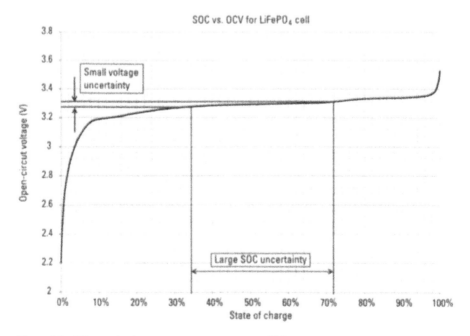

**Figure 6.1** Effects of voltage measurement error on SOC error.

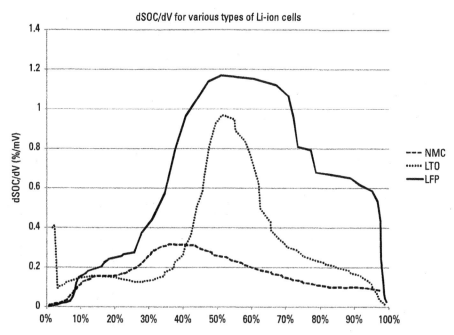

**Figure 6.2** $dSOC/dV$ versus SOC for different cell types.

failed and the system is permitted to continue operating, limits should be chosen such that worst-case errors do not lead to repeated overcharge and discharge events. If an actual measurement is not needed, the simplest strategy is to set an alarm signal from the measurement circuit that indicates a cell measurement has reached a certain limit. If it is not necessary to know which cell has reached the limit, these signals could be logical-ORed together to drive a common fault response.

Therefore, the specification for cell measurement accuracy is driven by the desired state of charge accuracy and the shape of the battery SOC/OCV curve.

Presently, cell measurement accuracies from off-the-shelf measurement ICs with a precision of approximately 1 mV and a total error of less than 10 mV are commonly available. This book will discuss the appropriate care in design required to ensure these levels of accuracy can be maintained in real-world implementation under all possible operating conditions.

If an asymmetric dual-measurement scheme is used, the assumption is that the performance of SOC, SOH, and limit algorithms will suffer if the primary measurement is lost. In systems that cannot accept this type of performance degradation, a triple-redundant primary measurement is warranted.

In many contexts, hazards associated with cell over-charge and over-discharge fall into the scope of *functional safety*. As such, the risks associated with overcharge and overdischarge must be enumerated and carefully analyzed. The analysis must include a discussion of the possible failure modes of the battery management system in which an overcharge or overdischarge hazard may be allowed to occur, as well as a number of other system-level factors. Many systems include multiple components that include a degree of protection (for example, the battery charger could limit the maximum charging voltage). The overall level of risk depends on many factors outside of the battery management system, including cell chemistry and design, battery pack architecture and construction and the specifics of the application. Specifics of design of battery management system functional safety will be discussed later, but the central guiding idea is that the battery management system is part of a complete battery system and that the entire response to an overcharge or discharge event should be considered for appropriate design.

Certain devices offer different levels of accuracy over different ranges. This is generally an acceptable and even desirable characteristic. A highly accurate voltage measurement is usually only needed when the cells remain within their safe operating area. If the battery cell voltages exceed the safe operating ranges, many systems will adopt a shutdown or limited operation strategy. In these modes, calculation of accurate SOC and power limits may not be required or even possible.

A particular concern is the response of the battery management system if a cell experiences a polarity reversal or a large overvoltage. Many semiconductor devices are not capable of handling negative input voltages and most circuits will have a maximum voltage that can be read on a particular measurement channel which is not that much greater than the highest cell voltage experienced under normal conditions. Damaged cell voltage measurement circuits may provide erroneous cell voltage readings that may not indicate a problem.

Many cell-stack measurement ICs offer the option to use an external voltage reference. This option can reduce measurement error. A standard safety feature should be verifying that the primary voltage reference is free of gross measurement errors which make the primary measurement unable to guarantee overcharge and overdischarge protection. Many voltage references have a dependence of output voltage on temperature.

Detailed analysis and testing of the cell measurement error should be performed on all systems.

Common errors include:

- Failing to account for voltage drop caused by drawing balancing current from cells across circuit components and connections. Systems demanding high accuracy with large balancing currents are the most susceptible.
- Failing to account for voltage drops caused by bus bar and cell interconnect impedances, which may vary from cell to cell.

- Not performing analysis and validation over the full range of voltages and temperatures expected. Many measurement circuits will have varying accuracy over these ranges.

As it is the method by which overcharge and overdischarge are prevented, the cell measurement system must be robust against failures where cell voltages are beyond the normal battery limits but values are reported to the software which appear normal. Depending on the components and measurement architecture chosen, the following possible failures should be considered:

- Is it possible that a multiplexed measurement device is not measuring all of the cells, measuring the same cell for more than one of its connected cells?
- Is it possible that a constant, but erroneous, cell voltage measurement is being reported?
- For parallel-addressed devices, are "stuck-at" faults for the chip select lines considered? These may cause the wrong device to report, leaving some cells unmeasured.

These types of failures are particularly dangerous. The battery may be severely overcharged or overdischarged if these events occur for long periods of time and remain undetected.

A redundant measurement architecture prevents a single-point failure from causing many of these conditions. Good design practice should be followed to minimize the number of multiple-point failures, leading to loss of both measurement circuits if failures with a common root cause are known.

## 6.2 Current Measurement

String current is the other fundamental battery quantity that a battery management system usually measures. As all cells are connected in series, a single current measurement will provide the current flowing in each of the cells.

For the same reasons as discussed above for voltage measurement, it may be required to perform multiple redundant current sensor measurements.

- If accurate current sensing is required to achieve satisfactory state of charge performance, and state of charge performance is a critical factor warranting robustness against a single-point failure, a second current sensor may be required.

- If there exists a significant range of currents that exceeds the battery limits but there is no other method likely to detect (such as extreme cell/pack voltages caused by high current flow) or prevent (such as passive overcurrent protection devices (i.e., fuses) these excessive currents, a second current sensor is warranted.
- If accurate reporting of the battery current is used by an external device for a safety critical function (i.e., the application cannot tolerate a battery management system current error), two current sensors should be used.

A number of sensor types are discussed next. Multiple sensors may be of the same or different types and may be used together. In the case of the second reason above, the accuracy to determine that the batteries are operating in a safe range is much lower than may be required for state of charge and other algorithms. An asymmetrical redundant approach would be sensible for this type of application. For reasons which will be discussed, multiple sensors are sometimes also used to provide a higher degree of accuracy.

When analyzing current sensor accuracy and reliability, the entire signal chain must be considered. This will usually consist of:

- The sensor or sensors themselves;
- The analog signal connection between the sensor and the measuring device;
- Analog prefilters;
- Amplifier (shunt signal only);
- Analog-to-digital converter;
- Digital filtering;
- Digital integration;
- Isolation barrier.

### 6.2.1 Current Sensors

Regardless of the type of current sensor selected, a number of fundamental parameters apply.

The range of the sensor must be large enough to cover the entire expected range of battery current in both charge and discharge, remembering that many battery systems may not have equal current capability in both directions. It is good practice to ensure there is adequate headroom in the sensor measurement range.

Many battery systems experience a wide dynamic range in the application currents. A typical automotive battery system may be discharged much more quickly than it is charged. Because the magnitude of certain types of measurement errors (including nonlinearity, discretization, offset) depends on the full scale of the current sensor used, it is often difficult to measure large and small currents with the same degree of relative accuracy. In some applications, this is not an issue because errors in small currents have a consequently much smaller effect; however, when small currents exist for a long period of time and current integration is used for SOC determination, errors in small currents can become significant.

The battery management system may measure the analog output of the sensor directly, or the sensor may contain the analog measurement electronics and interface with the battery management system using a digital interface. In the case of an analog sensor, the battery management system's analog measurement circuit must be designed to minimize measurement error.

The bandwidth and frequency response of the sensor are often overlooked. In many applications the battery current can change quickly and the sensor must be fast enough to capture dynamic current changes. The maximum expected slew rate (in both directions) should be known while designing the battery management system. The sensor, sampling rate, and associated circuitry are all affected by this. Often the amp-hour error associated with a current measurement that is slower than the fastest possible load slew rate can be shown to be small and generally equal in both positive and negative directions.

### 6.2.1.1 Shunts

A shunt is simply a precision resistor of a very low value. The shunt will have a voltage drop across it proportional to the battery current. The shunt resistance is small enough to make this voltage drop negligible in the context of the high-power current path of the battery system, but large enough to be measurable by the battery management system.

Shunts have a *four-wire* or *Kelvin connection* (see Figure 6.3). In this arrangement, the current carrying terminals and the voltage sensing terminals of the shunt are separated.

Shunts are inexpensive and available in a wide variety of ranges. Many shunts have a 50-mV or 100-mV output at the maximum measurement current. Their simplicity leads to very high reliability, and no external source of power is required.

Besides their low cost, shunts are generally highly accurate (accuracies of 0.1% to 0.5% are not uncommon). Shunts are available in a number of sizes, but shunts for very high currents can become very large and heavy.

The small shunt signals must be amplified before measurement and the circuit used to measure shunt voltage must be very high impedance. Shunt

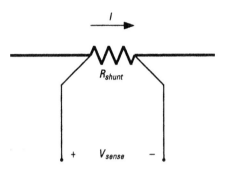

Figure 6.3  Kelvin connection for shunt.

amplifier circuits require a number of characteristics to achieve good measurement accuracy:

- *High common-mode rejection ratio:* The shunt amplifier circuit must reject both ac and dc common-mode errors.
- *Low dc offset:* Most battery systems must measure small currents that are both positive and negative. The dc offset can be particularly problematic and cause charging currents to be mistaken for discharging currents, and vice versa. Shunts themselves have the inherent advantage of not having any significant intrinsic dc offset.
- *Accurate, high gain:* Shunt amplifiers must frequently provide accurate gains of 100 or more. Gain errors translate into current measurement errors.
- *Good thermal stability:* Both the shunt and the amplifier circuit may operate at different temperatures, and the amplifier circuit may have gain and offset errors that depend on temperature.

In the case of systems that must operate over wide temperature ranges, the battery management system must consider the change in the resistance of the shunt as its temperature changes. A typical shunt (see Figure 6.4) may have a thermal coefficient of resistance of 80 ppm/°C. The shunt also self-heats due to battery currents.

Shunts have a very high measurement bandwidth because they are designed to be noninductive. As such, they can measure very rapid changes in battery current; however, it also means that conducted emissions from the load (common with load devices that contain power electronics) will travel unabated into the shunt amplification and measurement circuit. In most cases is it desirable to filter these out.

# Measurement 81

**Figure 6.4** Shunts for current measurement.

In the context of large-format systems, it is important to note that the shunt is referenced to the *voltage of the battery stack*. Although the voltage across the shunt itself is very small, the voltage between the shunt and other battery components could be very high. The electronics to process, amplify, and measure the shunt signal must also therefore be referenced to the battery stack. In applications in which isolation is required, the shunt measurement must be transferred across an isolation barrier.

The lack of isolation also creates a potential source of measurement error due to common mode voltages. When the load current increases, the common-mode voltage ($V_{s+} + V_{s-}/2$) increases by $\Delta I / R_{load}$. This should not create errors in the amplification of the differential-mode signal ($V_{s+} - V_{s-}$). The effect of this can be reduced by installing the shunt in the low side, where the common-mode voltage will be much smaller. This is only the case when the shunt amplifier and/or entire BMS is referenced to the negative terminal of the high-voltage battery. An isolated shunt amplifier also solves the problem.

### 6.2.1.2 Hall Effect Sensors

The Hall effect is the production of a voltage proportional to an electric current in the presence of a magnetic field.

Raw Hall effect devices have significant temperature dependence and other undesirable effects. Commercial Hall effect current sensors (see Figure 6.5) produce a voltage signal that is processed by an integrated signal processing circuit that removes a large portion of these errors.

Hall effect sensors have advantage that the signal voltage is isolated from the high-current path. In many sensors, the sensor has no high-current terminals

**Figure 6.5** Typical Hall effect current sensor. (Courtesy TKD.)

and a current-carrying bus-bar or cable is passed through the sensor body. The installation of Hall effect sensors is direction-dependent; the sensor will read negative values if it is installed incorrectly.

Hall effect sensors must be powered by an external source which in most cases will be the battery management system electronics. The current supply capability of the electronics must be verified. In some cases, this voltage source also defines the full-scale voltage for the measurement output (i.e., for a given current, the output voltage is a constant fraction of the supply voltage, rather than a constant voltage). This is known as a *ratiometric* sensor and prevents current measurement error due to different voltage references. The ratiometric sensing technique is usually preferred due to higher accuracy, and reduced need for a high-precision absolute reference. As the supply voltage also defines the scaling, it should be provided by a stable precision reference and used with a ratiometric analog-to-digital converter that accepts an external reference. Finally, the current supply capability of the voltage reference device must be adequate to drive the current sensor with sufficient headroom. Many precision reference devices are designed only to provide analog reference voltages with limited current capability.

Readings from Hall effect sensors can be disrupted by magnetic fields. The installation of the Hall effect sensor must protect it from external magnetic fields.

Open-loop and closed-loop types of the Hall effect sensor are available. The closed-loop type includes an extra coil around the magnetic core. By controlling the magnitude and direction of the current in this coil, the total flux in the core can be set to zero, and the magnitude of the current required to null the flux produced by the battery current is proportional to the battery current. Closed-loop sensors have higher accuracy and reaction time, and are less prone to magnetic saturation but are larger, more expensive and consume significantly more power (due to the requirement to drive the secondary coil). Both types find applications in large-format systems. Automotive applications tend to use open-loop sensors due to cost sensitivity.

Hall effect sensors are available with unipolar or bipolar outputs. Unipolar Hall effect sensors map the entire range of currents into a positive output voltage, and bipolar sensors have positive voltages for positive currents and negative voltages for negative currents. Bipolar sensors require bipolar supply voltage circuits that are generally required in very few other types of control circuits.

Additionally, the output of Hall effect sensors should have a load resistor attached to ensure that a minimum load condition (specified by the sensor manufacturer) is met. This is necessary to ensure stability of the output.

Most Hall effect sensors suffer from zero-offset error (sometimes called "electric offset"). It can be defined as the output of the sensor when there is zero current flowing. This offset is generally constant across the current measurement range and is best modeled by an additive constant error to the current measurement. This error can be positive or negative in sign. It varies with temperature and may vary from device to device and also over time with a single sensor. When current measurements are integrated for the purposes of evaluating battery state of charge, this type of error contributes significantly to measurement error. Because the error is of the same sign regardless of the direction of the current, there is no cancellation of errors during currents of opposite sign. If the current signal containing this type of constant additive error is integrated, the integration error increases linearly with time, and even small errors can become very large for systems with long operation cycles. A current sensor with this type of offset error equal to 0.001C (which is a very accurate sensor) in continuous duty will accumulate 16.8% SOC error in one week of operation.

For this reason, it is desirable, when possible, to use zero drift compensation and zero current cancellation with Hall effect sensors. A number of battery systems do not see continuous connection to their load devices. When the battery is disconnected by means of relays or contactors, regardless of the current measurement, the current is known necessarily to be zero. Therefore, the current signal should be set to zero inside the battery management system regardless of the sensor signal. Also, this provides an opportunity to measure the zero offset of the sensor and adjust subsequent measurements to use this new zero point. This value can be used over the next operation cycle to achieve better performance. This will provide some degree of temperature compensation as well as adaptive calibration to account for aging of the sensor.

Other types of error include hysteresis errors (caused by magnetization of the core), gain error (difference in the sensitivity of the sensor between the specified and actual value), and nonlinearity (deviation from a linear response between current and sensor output). Hysteresis and gain errors are usually symmetrical about the zero-current point (i.e., the sign of the error is the same as the sign of the current being measured). In this case, if the current profile of the battery is roughly symmetrical (equal lengths of time spent charging and discharging, with roughly equal magnitudes), the integrated errors will be equal in magnitude and opposite in sign and tend to cancel each other out. The degree of symmetry expected in the current profile should be considered when analyzing current sensor errors. Gain errors are also proportional to the current flowing. Systems that do not operate often at peak battery current are more tolerant to gain errors in current sensing.

Examples of symmetrical current profiles would include: charge-sustaining drive cycles in a hybrid-electric vehicle, frequency regulation, or peak shaving in energy storage applications.

Examples of nonsymmetrical current profiles would include battery electric vehicles or backup power systems that are usually charged much more slowly than they are discharged.

### 6.2.1.3 Magnetostrictive Sensors

Magnetostrictive sensors are a relatively new technology compared to the other two options discussed earlier. These sensors operate on the principle of magnetostriction; a current creates a magnetic field that creates strain in a material. The strain is measured with a strain gage, and is related to the current.

### 6.2.2 Current Sense Measurement

A great deal of emphasis is placed on the resolution of the analog-to-digital converter used to convert the current sensor measurement to a digital signal. Although this is an important attribute, there are a number of upstream decisions that can affect the real-world utility of this measurement.

- *Connection to the battery management system:* In the case of a shunt, the current sense signal is low voltage and it must be connected to a high-impedance input. This type of connection is very sensitive to noise and should be shielded with appropriate shielding and grounding techniques.
- Use twisted-bundle connections to minimize current-mode interference.
- Determine the bandwidth and sample rate of the converter and sensor as well as the expected frequency content of the load current. An analog prefilter should be used to remove frequency content above the Nyquist frequency of the conversion circuit.

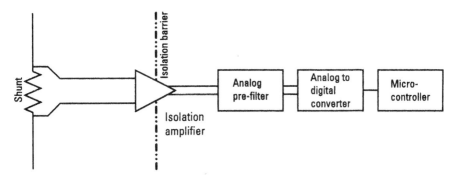

**Figure 6.6** Block diagram of current measurement circuit.

- Active filters must be designed for zero dc offset. It is easy to go to great lengths to minimize offset errors in the transducer, only to reintroduce the identical problem by adding dc offset in the filter and amplifier stages of the measurement circuit.
- Consider a separate precision voltage reference for the current sense circuit.
- If possible, use a ratiometric sensor and converter to eliminate voltage reference errors.
- The measurement circuitry, especially when used with a shunt, must offer high common mode rejection ratio. Locate shunts appropriately to reduce common-mode voltages.

Current sensors may be internal or external to the battery management system modules. If they are internal, the entire battery current must flow through the battery management system component. With high voltages and large currents it may be inefficient to use an internal sensor due to the PCB space required for large current traces and high-voltage separations. Otherwise, small PCB-mount sensors and shunts exist and can provide a very compact package.

The problem of maintaining good integration accuracy with wide dynamic range and long integration times is sometimes addressed through the use of multiple sensors with different measurement ranges. Hall effect sensors with two or more ranges are commercially available. The smaller range simply saturates when its maximum measurement range is exceeded and the larger range is then used. Careful attention should be paid to the switchover point between the low-range and high-range sensors. Figure 6.6 shows a typical inteface circuit for a current shunt. Hall effect interface circuits are similar.

## 6.3 Synchronization of Current and Voltage

Accuracy and sample rate of current and voltage measurements are important individually, but it is also important to understand the time-domain relationship between both measurements.

For reasons to be developed more fully later, being able to obtain synchronized or coordinated measurements of voltage and current is important for many advanced features of large-format battery management systems.

Measuring cell impedance requires determining the instantaneous relationship between voltage and current, and therefore the current and voltage sampling must be somehow coordinated.

All measurement devices therefore must be deterministic in time. This problem can be stated as follows: the device must accept an external trigger signal to make measurements, there must be a minimum time for the measurement to begin, and there must be a maximum time for the measurement to end. The total width of the measurement window must be short enough that the quantity being measured can be approximated to be constant during this period. The minimum time to begin the measurement must be short enough that the system's real-time characteristics are not degraded.

A number of methods can be used for triggering a measurement. The simplest is a digital logic signal that uses a level or edge to start the measurement timing. Some devices will begin converting on the final edge of the final bit of a command message in a serial communication stream which is controlling the device. If this message has been transmitted asynchronously using interrupts (common for buses like $I^2C$ and CAN), the time at which this edge occurs may be difficult to determine as it does not occur at a given point in normal program execution. A detailed timing analysis may be required.

The simplest approach conceptually is to synchronize all measurements very tightly in time such that the difference can be assumed to be negligible and all measurements are assumed to be simultaneous. This approach provides for the simplest software solution; however, this implementation may be more challenging in hardware. A number of cell-measurement IC suppliers offer ICs which perform all cell measurements within a few microseconds, which for most systems can be considered to be simultaneous. These chips may use an analog-to-digital converter for each cell to allow for synchronized measurement, which may increase their cost, or may use a faster ADC which is multiplexed between cells, but this may limit the accuracy available when trying to perform measurements fast enough to consider them synchronous.

A further level of complexity is created when multiple measurement ICs are used on a single module, and when slave modules are used with a single master in a distributed architecture, leading to synchronization issues between ICs and slave modules.

Another possible solution is to attempt to coordinate measurements without achieving complete synchronization. If voltage and current measurements occur at different times but with constant offsets, it may be possible to interpolate data to obtain voltage or current data at intermediate time points.

A third possibility is to allow voltage and current measurements to proceed asynchronously. Because many of the measurements that require synchronicity of these two measurements do not need to be made continuously, a type of opportunity calculation can be made on individual cells when a current and voltage measurement occur at the same time. This requires no special work in hardware or software to synchronize the measurements; however, careful postprocessing of the data is required to identify where a synchronized measurement has been made. This may not occur with great enough frequency to provide enough fidelity for accurate battery state determination, but it provides for a simple implementation.

Measurement voltages are usually transmitted over a communications bus in packets or messages of fixed length. The transmission will occur after a number of messages have been made, and may occur at a nondeterministic time interval after completing the measurements. Therefore, the time at which the message arrives at the receiving device is not useful to determine when the measurements were actually made. A good strategy is to attach a timestamp to the measurements made by slave devices. A master clock can be started on the master device with periodic synchronization of all slave devices to the master. It is not necessary that this be an absolute time reference nor does it need to be monotonic; a rolling counter with a large enough period is adequate. If all measurements are made simultaneously, or within a very short time period, one time stamp may suffice. If measurements occur at regular intervals, a start time (and possibly measurement interval) may be used. The master clock and timestamps can be used to synchronize voltage and current measurements made on different devices to ensure accurate battery state estimation.

## 6.4 Temperature Measurement

Battery performance and behavior have a great deal of temperature dependence. In addition, preventing operation outside of the battery's safe temperature region is vital. For these reasons, most battery management systems will incorporate one or more temperature measurements.

The range of the temperature measurements should cover the widest possible operating range expected in the battery system, plus enough margin to account for possible errors in measurement at the limits of the operational range. Most lithium-ion batteries have a minimum operational temperature between

−10°C and −30°C and a maximum operational temperature between +45°C and +65°C.

A number of different types of sensors can be used to measure temperature.

Thermocouples generate a very small voltage across the junction of two dissimilar metals that is dependent on the temperature difference between the junction and the body of the metals. Thermocouples are generally useful for measuring large temperatures and are often found in laboratory applications. Use in an embedded system is less common due to the very small signals and low noise immunity. Thermocouples are also not well adapted for measurement over temperature ranges that are appropriate for lithium ion batteries. For this reason, other types of temperature sensors are more common.

Thermistors are widely used for temperature measurement with embedded control systems in many industries, including automotive. Thermistors have a resistance value which changes with temperature (see Figure 6.7). They are available with different nominal resistances (usually quoted at 25°C), different temperature dependence coefficients and in either negative temperature coefficient (NTC) or positive temperature coefficient (PTC) types. NTC thermistors increase in resistance with decreasing temperature and PTC thermistors increase in resistance with increasing temperature.

The temperature coefficient should be selected to optimize the thermistor for the measurement ranges of interest. Most lithium ion batteries will not

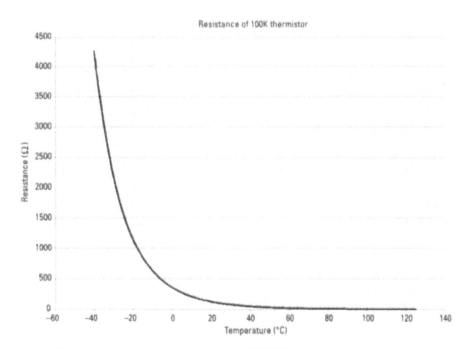

**Figure 6.7** Resistance versus temperature of a 100K NTC thermistor.

operate below −20°C to −30°C. Temperatures above 50°C–65°C are widely considered excessive. Temperatures outside of this range usually do not require a great deal of precision because battery operation is inhibited. Below temperatures around 0°C, many cells cannot be charged without the possibility of lithium plating. At temperatures below 25°C, charge and discharge performance may be derated due to increased battery impedance. Therefore, optimizing the measurement circuit to provide best accuracy in a range of approximately −5°C to 30°C would be suitable for most cell types.

The relationship between thermistor resistance and temperature is complex and nonlinear over the range of temperature of interest. In many applications, the resistance of the thermistor is approximated as a linear function of temperature. This can produce significant losses in accuracy. A number of nonlinear models of thermistor resistance exist, but for most battery management system applications, a simple one-dimensional lookup table with linear interpolation can be used to provide acceptable accuracy without having to perform complex calculations or rely on numerical models.

The consequences of selecting NTC or PTC thermistors, apart from the standard engineering constraints of cost, performance, and size are influenced by the way in which extreme values of temperature are interpreted. Due to the possibility for lithium batteries to go into thermal runaway and generate significant heat, it is more likely to encounter (and more important to detect) very high temperature excursions than very low temperatures.

The reference voltage of the thermistors is another consideration. Because the thermistors are located very close to the cells and it is often common to route the voltage and temperature sense wiring in the same conduits and harnesses, a possibility exists to have short circuits between temperature sensing circuits and cell voltage. If the cells are isolated from earth ground and a short of this type occurs, an isolation fault will be created. If multiple shorts of this type occur (common with harness pinching and other process errors), a short across several cells can be created. If the temperature measurements are referenced to the lowest cell in the battery stack, very high potentials could exist between the temperature circuits and the most positive cells. A desirable compromise is to group temperature measurements of cells at similar potential to be referenced to the lowest potential in the group of cells. The potential difference between the cell stack and the temperature sense circuits that are most likely to experience a short circuit to the stack can be minimized, and straightforward techniques can be used to ensure that the worst-case short circuit will not damage the battery management system or lead to a high-energy circuit.

Measuring the resistance of a thermistor usually is accomplished by creating a voltage divider using the thermistor and one or more fixed resistors. A fixed voltage can then be applied to the entire series combination of resistors and a measurement made of the voltage across the thermistor, which will be proportional to its resistance.

To prevent a situation in which a harness short circuit will cause a short circuit of the thermistor supply voltage, a fixed resistor can be placed on both sides of the thermistor. In this situation, a shorted sense circuit will not cause damage to the measuring circuit.

Worst-case analysis of thermistor accuracy is more complicated than it may initially appear. The nominal resistance of the thermistor has a basic tolerance band. The temperature coefficient (usually denoted as $\beta$) also has a tolerance factor, and $\beta$ itself may depend on temperature. This is a deterministic effect that can be compensated out using an appropriate lookup table as discussed earlier. Tolerances also apply to the fixed resistors used to create the voltage divider. Due to the nonlinear relationship of temperature and resistance, the accuracy of the circuit will usually vary significantly over the range of interest.

If thermistors are sized improperly such that a significant current flows during operation, they may self-heat, causing erroneous readings. Thermistor sensing circuits should ensure that thermistor current is kept to a minimum so that self-heating is negligible.

Temperature measurement accuracy has a number of elements, a number of which lie outside of the electronics and sensor signal chain. Remember that the sensor is measuring the temperature at the sensor but it is the temperature inside the cells that the battery management system is trying to determine. Sensor installation affects the thermal resistance between the cell enclosures and the sensor, leading to measurement errors. Good, consistent contact is required for accurate measurement, but this may be in conflict with the need for electrical isolation.

Depending on the size of the cell, the cell itself may experience a thermal gradient, especially with systems using active heating and cooling. Calculating the *Biot number* for the cell using a numerical model is an effective method of determining whether this is a concern. Biot numbers below 0.1 indicate that all areas of the cell can be assumed to be at equal temperature. This can be calculated using numerical methods, computer simulation or empirical equations.

Temperature sensor location is also an important factor. In a large battery system, temperature gradients are usually expected. If the system operates with wide gradients or operates close to the edge of the safe operating area, it is possible that individual cells whose temperature is not measured can exceed the safe operating area (SOA) while all temperature measurements remain inside the acceptable range. Additionally, if a cell experiences a spontaneous thermal event, the further the sensors are located from this cell, the lower the probability that the high temperature will be detected, and the longer the detection will take.

The battery management system designer must be aware of these thermal limitations and a multidisciplinary approach is necessary to achieve a satisfactory result. A systematic approach to validation is necessary. The following concerns are important in the design of temperature measurements.

- What is the largest error that could be expected between the hottest/coolest measured cells and the actual hottest/coolest cells in the battery system? This should be verified at the extremes of battery, ambient temperature load and thermal management input (where the gradients will be the highest).
- Under maximum thermal load, how much does the measured temperature lag the cell core temperature? Is it possible that a cell's internal temperature exceeds the safe operating area before the temperature measurements indicate there is a problem? If this situation exists, it may be possible to improve the performance by incorporating a simple model of heat generation and rejection into the battery management system. This model can predict the cell internal temperature given battery current, cell resistance, heat rejection, and heat capacity. Long-term accuracy of such a model will certainly not be adequate to eliminate the need for temperature sensors, but under heavy thermal transients, it could provide a more accurate estimate of peak cell temperature than measurement data.
- If a cell experiences thermal runaway, how much time is required, in the worst case, before an abnormally high temperature is detected?

It is acceptable to have a measurement accuracy that varies over the range of temperature being measured. At high and low temperatures, battery performance is already limited or even prevented altogether and high temperature fidelity does not provide much benefit.

An important case to consider is where a single cell (not necessarily even a single series element) experiences an internal short circuit and begins to increase in temperature or even enters thermal runaway. If this type of condition can be detected, a number of countermeasures could be activated. In most applications, placing a temperature sensor on each individual cell is out of the question due to cost and complexity, but if determining the accuracy and location of the particular cell are unimportant, a number of less costly methods are possible. In many cases it may not be critical to know which cell has reached a very high temperature, only to know that it has occurred.

Inexpensive PTC elements that have a significant resistance increase around a threshold temperature can be used to detect a high temperature cell. Fusible wires with a melting point tuned to the temperature that corresponds to the threshold of overtemperature, which could also be used. Detecting an open circuit on the wire indicates that a cell has reached this temperature.

A technique using closely matched diodes installed in parallel can also be used to provide inexpensive maximum temperature detection. The voltage drop across a forward biased diode is a function of the temperature of the junction, according to the equation. If a number of diodes are placed in parallel, the lowest voltage drop will be across the warmest diode, allowing the highest temperature to be measured with a very inexpensive set of sensors.

This method is potentially useful for hot-spot detection but has a number of disadvantages, notably the very small difference in diode voltage between temperatures of interest, requiring accurate measurement and careful attention to other voltage drops in the circuit due to connectors, traces, and other components of the conductive path.

## 6.5 Measurement Uncertainty and Battery Management System Performance

All measurements in engineering systems contain some degree of uncertainty. The maximum acceptable error should be determined as part of the system requirements. As part of the implementation and validation process, a worst-case circuit analysis should be performed using the specified tolerances and limits of the components. The output of the worst-case circuit analysis should provide a complete engineering analysis of the capability of the designed hardware. This should be performed along with testing of a number of sample devices, ideally over the expected operating range of temperature and other environmental variables, to characterize what typical measurement errors will be. Armed with knowledge of both typical and worst-case measurement errors, system designers can set about to ensure that the effects of measurement errors do not unacceptably compromise battery management system performance.

## 6.6 Interlock Status

Many battery systems operating at hazardous voltages employ an interlock system. The purpose of the interlock is to ensure that barriers that separate persons from conductive parts (wires, bus bars, terminals) at hazardous electrical potentials are intact before energizing the conductive parts. Interlock functionality is critical to minimizing the risk of an electric shock hazard. As the battery is a source of electrical energy (but often not the only source), the battery management system often monitors the state of interlocks to determine if actions should be taken or inhibited. Interlock loops are designed in such a way that a conductive path between the source and the detector (sink) of the loop signal is interrupted if an insulating barrier is compromised. Examples of how this may be done include:

- Switches are activated by the removal of doors, covers, or panels, which provide access to current-carrying components.
- Additional low voltage terminals in high-voltage connectors are disconnected when the HV connector is unmated. These terminals should be of a break-first/make-last design.

The interlock loop should protect all reasonable points of access to hazardous voltages by passing through all devices and interconnections.

The operation of the interlock loop is simple in principle but due to its critical nature, care should be taken to avoid malfunctions which could lead to the battery system being inoperable (in the case of a false "open") or in reduced protection to personnel (in the case of a false "intact").

Loops may have a single source and sink or they may use multiple sources and sinks. If the loop uses a single source and sink, these could be placed on the same device or on different devices. Placing the source and sink separately offers the shortest and simplest path through all of the components but introduces a number of complications. Detection thresholds for an intact loop can be tightened if the source signal is known. If the same device acts as the source and sink, source measurement is significantly easier.

# 7

# Control

## 7.1 Contactor Control

In many systems, contactor control is a vital function that must be implemented with diligence. The contactors (see Figure 7.1) and associated disconnect hardware form the last line of defense for protecting the battery system, and they also are used every time the battery system is brought into service. If the contactors do not function correctly and the battery cannot be disconnected, an important method of preventing overcharge and overdischarge is lost.

Most battery management systems are required to diagnose contactor defects, including contactors that have failed to open (under certain circumstances the contacts can weld) and those that have failed to close. Certain contactors require different current levels for pull-in and hold, and the "economizing" function to reduce current after closing is often placed in the battery management system.

Contactors and relays are electromechanical switches in which a solenoid (coil) is energized by a low-power circuit to mechanically close contacts of a higher power circuit. Compared to solid-state semiconductor switches, contactors provide reliable isolation due to physical separation of the primary and secondary circuits, high levels of amplification (very small coil drive power can be used to switch very high currents and voltages) and do not require level shifting and gate drive circuitry. They provide very low resistance to current when closed and very high resistance when open, unlike semiconductor switches, which always have some leakage current and which usually have a higher on-resistance.

Contactors for dc circuits are specially designed to quench the arc that may form when the circuit is opened, especially with an inductive load. Contactors are generally highly reliable, but they do have a few susceptibilities of which system designers must be aware. Because the primary function of a contactor system is to connect and disconnect the battery to/from the load when desired, the most concerning failure modes are contactors that fail to close and those that fail to open.

Closing a contactor into a short circuit will in many cases contribute to the welding of the secondary contacts. This can occur with capacitive loads, which create a large in-rush current when the contractor is closed.

Welding of contactors can also occur if the contactors are exposed to higher than rated current, causing the contacts to be forced apart while the solenoid coil is still energized, which promptly forces the contacts to close again.

An unstable control circuit can cause rapid closing and opening (sometimes called "chatter") of the contactor coil. The bouncing of the contacts against one another often leads to contact welding. Contacts can also weld if the coil current is not brought to zero quickly enough, creating a "soft" opening condition where the contacts can reclose during the opening sequence.

Contactors that are welded and fail to open can create a potentially hazardous situation. It may become impossible to disconnect the battery from the load, meaning the current cannot be cut off and overcharge and overdischarge hazards can no longer be prevented by the battery management system. Connections that are normally expected to be in a safe condition when the battery is disconnected may remain electrically live at all times, leading to the potential for electrical shock and fire hazards.

Equally problematic, although less hazardous, are contactors that fail to close. This will prevent connection of the battery to the load and make the system unusable.

Contactors may be damaged in this way by overload (either long-term overload or short circuit). Overvoltage will also lead to excessive arc energy and can lead to contact failure. High temperature can lead to contactors that fail to close due to thermal damage to the armature.

All contactors have maximum rated lifetimes. These lifetimes will consider the maximum number of cycles that the contactor can endure under various opening conditions. Opening contactors at high currents will drastically reduce their available life.

Methods of addressing all of these potential causes of contactor damage will be addressed.

## 7.2 Soft Start or Precharge Circuits

Many battery management systems are required to control a soft-start or precharge circuit to allow the battery to be connected to a large capacitive load. If the battery is connected directly to an uncharged capacitive load, the in-rush current will be limited only by the internal impedance of the battery, load, and interconnect, which will generally not be enough to prevent very large, potentially damaging currents.

This type of uncontrolled soft start is likely to lead to welding of contactor contacts as the contacts close across a large voltage and experience a very high current. Contactors rated for seemingly high currents and operating voltage can be susceptible to this type of damage at even modest voltages across the contacts during closing. Therefore, contactors should be closed across small voltages and current should not flow until the contacts are fully closed.

The most common method (see Figure 7.2) of addressing this problem is to include a soft-start circuit, comprising a resistor in series with an additional relay or contactor, installed in parallel across one of the main contactors.

When the battery is connected to the load, appropriate contactors are closed to create a connection to the load through the precharge resistor, limiting the current to $V_{batt}/R_{precharge}$, and causing the voltage on the capacitive load to rise exponentially. When the voltage on the load is high enough that the voltage across the main contactor is negligible, the main contactor can be closed, effectively shorting out the precharge circuit. The precharge contactor can then be opened and the battery is connected to the load and ready for operation.

The most basic method of ensuring that the precharge is complete is to simply time the sequence of the contactor activations so that the precharge

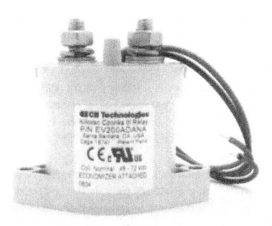

**Figure 7.1** Large-format battery contactor. (Courtesy Tyco Electronics Corporation.)

circuit is closed for an appropriate length of time to allow the load capacitors to fully charge before closing the main contactor. Although straightforward, this method has a number of drawbacks in that it does not allow for any fault detection or diagnosis, nor does it potentially account for unexpected changes in the resistance, capacitance, or leakage current across the load. As well, the voltage differential across the main contactor at a given time after closing the precharge contactor will depend on the battery voltage. Any of these effects (some of which occur naturally as components age) could cause the main contactor to close into a large inrush current with potentially disastrous results. Therefore, more care is usually warranted for this function.

The most direct method is to simply measure the voltage of the load and compare it to the battery voltage. This will determine the voltage across the contactor terminals. The main contactor should be closed when the voltage difference between the battery voltage and load voltage is small enough to operate the contactor without any damage. This will account for normal variation in battery voltage, a number of possible faults such as changes in load capacitance, and precharge resistance, and handle small conductances across the load capacitor.

If a large current is drawn by the load during the precharge sequence, the load capacitor will either never charge completely or charge very slowly. Any current drawn during the precharge will flow through the precharge resistor, which is usually sized only for a surge application and not capable of handling significant load currents for more than the brief period of time during the precharge. A good precharge control sequence will verify that the load voltage is rising at the correct rate and open the precharge contactor to abort the sequence if the load voltage rises too slowly or does not reach the required voltage quickly enough. It may also be important to limit the duty cycle seen by the precharge resistor by preventing multiple precharge attempts in rapid succession. These controls will protect the precharge resistor from damage due to overheating. Figure 7.3 shows voltage, current, and power during the precharge sequence.

The value of the load capacitance may vary significantly. Many capacitors have wide tolerance bands especially when device aging and temperature effects are considered. Additionally, many devices may be required to perform a "hot start," wherein the battery is expected to reconnect to the load before the load capacitance has been fully discharged. In this case the acceptable range of voltage/time relationships will need to be modified to account for the partially charged load.

If the load voltage increases very quickly, it may be an indication that no load is connected or that the battery may be being operated in an unintended way. In some applications this may be undesirable and should therefore also end the precharge sequence.

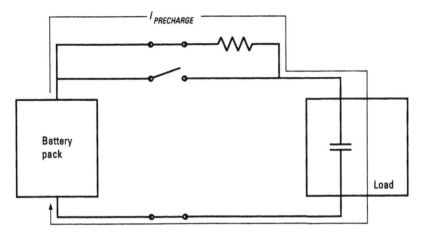

Soft-start or precharge circuit avoids large inrush current into discharged capacitive load when battery is connected.

**Figure 7.2** Soft-start circuit schematic.

## 7.3 Control Topologies

For reasons previously discussed, it is imperative that the battery system is capable of disconnecting the battery from the load to prevent a dangerous condition. Switching devices used to control contactors can fail in different modes including open or short circuit or anywhere in between.

The principal failure modes that should be addressed in the development of schematics and layout for contactor control circuits are:

- Prevent failures that cause contactors that should be open to close.
- Prevent failures that cause contactors that should be closed to open.
- Prevent failures that cause connection of the battery across a discharged capacitive load.

The safety requirements and level of hazard associated with unintended contactor operation will drive the level of care required for each of the above failure modes. Next methods of implementation will be discussed.

Placing multiple switches in series is an effective method of increasing the reliability of interrupting the contactor coil current and therefore opening the contactor; however, this is at the expense of reducing the reliability of ensuring that a contactor remains closed as there are now two switches that could fail in the open condition. This type of trade-off is an important consideration in the implementation of safety critical functionality (see Chapter 22).

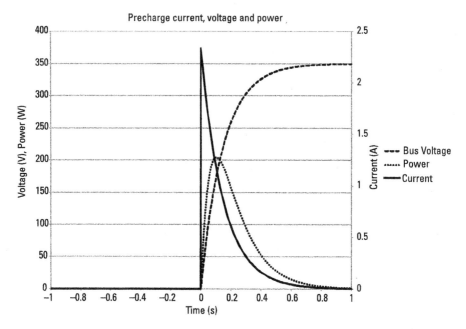

**Figure 7.3** Precharge sequence current, voltage, and power.

In most cases, the safety hazards associated with contactors that cannot be opened are considered to be more severe than the loss of functionality that occurs with contactors that cannot be closed. This leads to a fail-safe strategy in which multiple switches are placed in series that must be operating correctly for the contactor to close. This, combined with multiple contactors, creates a high level of robustness in the disconnect capability.

A common implementation (see Figure 7.4) places one switching device on the positive "high" side and another on the negative "low" side of the contactor coil. This has a number of advantages. Single shorts to ground or to contactor control power can no longer cause a contactor to close. The two devices are now likely of different design (complementary NMOS/PMOS transistors), which reduces the likelihood of a common root cause leading to multiple failure, as could be the case with two identical devices placed in series.

As it is common that multiple contactors need to be controlled, complexity and cost can be reduced by using a single device in either the high-side or low-side position to provide control to all contactors, in conjunction with individual devices in the other position to control each individual contactor.

In addition to the importance of this function, contactor control is generally one of the larger loads that must be controlled by the battery management system due to the magnitude of contactor pull-in and hold currents.

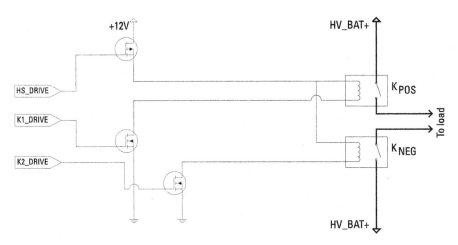

**Figure 7.4** Dual contactor drive circuit schematic.

The use of so-called smart semiconductor switches, commonly referred to as *high side drivers,* or *low side drivers,* for contactor control is warranted for many applications. These switches include protection against short circuits, over-temperature, electrostatic discharge (ESD), and overvoltage, and may be capable of dissipating the inductive switching load without external components. These types of faults in semiconductor switches may lead to failures in the short-circuit condition.

## 7.4 Contactor Opening Transients

The electromechanical nature of relays and contactors implies they are inductive loads with a significant energy storage in their internal magnetic circuit when the contactor is closed. This energy can damage switching electronics if not properly dissipated and can increase contactor opening time. The increased opening time is damaging to the contactor due to the longer duration of the arc and increased contact erosion.

The use of a standard freewheeling diode will protect the switching electronics from damage, but the energy in the contactor coil is dissipated only by the internal resistance of the contactor coil, diode and associated interconnections. This effectively results in an RL (resistor-inductor) circuit with an exponential decay of the current and therefore will lengthen the contactor opening time significantly. The diode should have a low forward voltage drop and fast but "soft" reverse voltage recovery characteristics. A Schottky diode is usually the ideal choice for this task.

A resistor can be placed in series with the freewheeling diode to reduce the time constant of the circuit. The resistor will dissipate the contactor coil energy and should have an appropriate power/energy rating for this application. Because it is a surge application, matching the resistor's continuous power rating to the peak power expected in the resistor may result in a very large device; however, if the resistor is unable to handle continuous duty, care must be taken to ensure that repeated contactor cycling does not damage the resistor.

A better option is the use of a transient voltage surge suppression (TVS) or Zener diode in the freewheel circuit. The TVS/Zener diode will allow a certain amount of back electromotive force (EMF), but will conduct at a defined forward voltage drop and clamp the voltage at this level. This changes the characteristic of the contactor opening. Because the voltage across the inductor is equal to the roughly constant forward voltage drop across the diode, and $V_L = L \, di/dt$, the current will decrease in a roughly linear ramp. This provides the fastest contactor opening. The voltage can be tuned by the selection of the TVS/Zener diode to provide a fast opening time without exceeding the voltage rating of the switching components. Use of a bidirectional TVS (see Figure 7.5) can also provide ESD protection for the contactor control outputs.

## 7.5 Chatter Detection

Contactor "chatter," or repeated opening and closing of the contactor armature, can occur for a number of reasons, but it nearly always leads to contactor damage including armature overheating due to excessive inrush current and contactor welding, creating an unsafe condition. Chatter should be prevented using a robust method to ensure this does not occur.

Chattering occurs due to dynamic interactions between the load profile presented by the contactor and the power supply activating it.

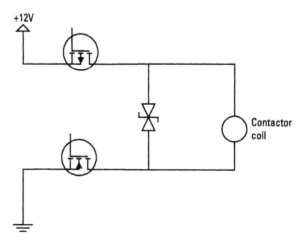

**Figure 7.5** Contactor EMF suppression with a TVS diode.

A low control supply voltage can cause contactor chatter. The minimum voltage required to ensure contactor closing should be provided by the contactor supplier and verified by testing under the full range of operating temperatures expected. A margin of safety should be added to account for error in the measurement of the contactor control voltage. A hysteresis band is also recommended to ensure that the voltage rises to an acceptable level for a sufficient period of time to prevent contactor actuation in a marginal condition.

Another possible cause of contactor chatter (even if the supply voltage is adequate) is a high-impedance current path through the contactor coil. This can happen due to malfunctioning semiconductor switches, poorly sized conductors, wiring harness or connector defects, or a high-impedance power supply. This can create an adequate voltage to activate the contactor, which dips rapidly due to the contactor inrush current, which causes the contactor to open, reducing the inrush current and restoring the power supply voltage, leading to oscillation.

If the contactors selected operate at the full control power supply voltage, a number of precautions must be taken. First, the acceptable supply voltage range must be restricted to ensure that adequate voltage is supplied to the battery management system before contactors are actuated. Second, an analysis of the voltage drops that exist in the contactor control circuit must be performed. This is typically dominated by the on-resistance of the high and low side control switches, but could include connectors, PCB traces, and a current sensing resistor, if used. This analysis must show that the contactors will be connected to a high enough voltage to close them properly. In the case where multiple contactors share some components of the drive circuit, consider that the worst-case condition must include the worst-case combination of hold and pull-in currents. It is often good practice to limit the number of contactors that close simultaneously to reduce peak currents and voltage drop and reduce the risk of chatter. Contactor activation can be staggered so that the high inrush currents do not overlap with minimal impact to battery functionality.

Using a contactor that has a lower rated voltage than the nominal supply voltage allows the use of a simple buck converter or PWM drive circuit to ensure that the lowest possible supply voltage can still supply the contactor adequately. An alternative is the use of a buck-boost dc-dc converter. This can prevent chatter or contactor failure to actuate due to low voltage, but adds cost and complexity. Both of these solutions add the risk of increased electromagnetic interference.

Some contactors can be chattered mechanically due to application of physical vibration or shock causing the armature of the contactor to be mechanically forced open. The contactor usually then closes again immediately. If there is a voltage difference across the contacts, this very often leads to contact welding. Damaged contactors may have other internal mechanical or electrical reasons that cause chattering. In these situations, detection of the current transient associated with closing can prevent permanent damage or dangerous situations.

For contactors with auxiliary contacts, chattering may be detected if the auxiliary contacts display intermittent connectivity. This may occur due to improper wiring or other system faults and lead to incorrect diagnosis.

Measuring coil current dynamically is an effective method to prevent contactor chattering. If the in-rush current associated with a contactor reclosing is detected due to any of the above reasons, the contactor can be switched off before the chattering has a chance to cause significant damage. For this to be effective, the current should be measured with enough accuracy to determine that the in-rush has occurred, and should be sampled frequently enough that the entire contactor closing event can be analyzed dynamically. A minimum of 10 sample points should be taken throughout the duration of the closing sequence.

The hold current of a contactor will vary with contactor temperature, which will be influenced by ambient temperature and in most cases, self-heating over time due to the contactor coil current. This requires caution if a simple maximum threshold is used to detect a chatter transient. It may be necessary to use a dynamically adjusted detection window that adapts for change in the steady-state current.

The closing transient should only occur in response to a command to close a contactor. If the closing transient is detected at another time, there is a possibility of contactor chatter. Opening the contactor drive circuit will minimize or prevent chattering. However, if the chatter detection circuit trips erroneously, the contactors will be opened while the battery system is in operation: a highly undesirable result.

## 7.6 Economizers

Most electromechanical relays require a significantly higher current to close the contactor than is required to hold the contacts closed. This is due to both the inductive effect of the contactor coil as well as the transient electromechanical effects that occur as the armature inside the contactor moves. A circuit known as an economizer (sometimes a coil economizer) is used to reduce the current supplier to the contactor coil after the contactor has closed. In some cases economization is optional to reduce power consumption and improve contactor reliability due to reduced heating, and in other cases it is required to prevent contactor damage from thermal stress.

A simple chopping strategy can be used to economize a contactor by operating the switching transistor in a pulse-width modulated mode. The simplest approach is to use an open-loop PWM control to provide a variable current supply to the contactor coil. A straightforward timing sequence can be used to control the current as the contactor closes. This strategy however does not account for variation in contactor current or for temperature-dependent contactor closing requirements—as the coil resistance increases, a higher voltage is needed to achieve pull-in and hold contactors closed. Contactors that are overeconomized may not have adequate holding force and open under vibration, whereas contactors that are undereconomized may overheat.

Economizers must also not interfere with the rapid dropout requirement to allow contactors to open quickly.

PWM economizers are a potential source of electromagnetic interference. The frequency components of the PWM signal will dictate the frequencies and magnitudes of radiated and conducted emissions. This involves not only the fundamental (switching frequency) but also the edge rates of the PWM signal. Fast edge rates have high-frequency components and will increase both the magnitude and overall frequency range where emissions occur.

If economizers are required to prevent contactor overheating, the contactor coil current measurement strategy is effective to detect contactors that fail to economize as well as other economizer faults.

## 7.7 Contactor Topologies

Many battery systems involve the use of multiple contactors for improving safety and functionality. Because the reaction to the most severe faults detected by the battery management system will be to open contactors to protect the battery system, a reliable disconnect means is vital.

The minimum disconnect topology would be a single contactor installed in the battery pack (see Figure 7.6). The contactor could be located at the positive, negative, or mid-pack positions. This will provide disconnection of the battery, but will not completely isolate the battery cells from the load. If the contactor becomes welded, the battery will remain permanently connected to the load with no secondary means of disconnection. No soft-start capability is

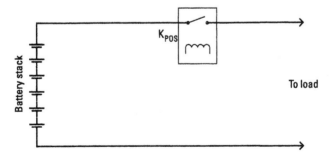

**Figure 7.6** Single positive contactor.

possible with this layout. This layout offers the lowest cost but also the lowest level of safety and functionality. If the battery system is referenced to earth ground at its lowest potential, using a single contactor on the positive terminal is a safer option as the negative terminal does not represent a significant voltage hazard.

Many battery systems use a pair of contactors (see Figure 7.7), at both battery terminals, especially in the case of a battery stack that is fully isolated from earth ground (as in most electric vehicles). This offers complete isolation from the load and from potential ground faults when both contactors open. If one contactor becomes welded, the second contactor can provide an effective means of isolation.

A contactor can be located at the mid-pack position. If the battery stack is split into two halves, a pair of mid-pack contactors (which may be actuated together or separately) can be used to provide effective disconnect means for sub-packs.

An additional contactor is often added for soft-start functionality (see Figure 7.8). This contactor is installed with a resistor in a parallel configuration to one of the main contactors. This contactor is usually smaller and may present different transient characteristics on closing. The precharge/soft-start contactor can be located on the positive or negative leg of the battery system.

## 7.8 Contactor Fault Detection

The contactor faults discussed earlier, included welded contacts and failure to close, have the potential to create dangerous conditions and detection methods are usually required to mitigate the risk associated with them.

There are two strategies principally used to determine the position of the contactors. One is to perform measurements of the actual high-voltage system

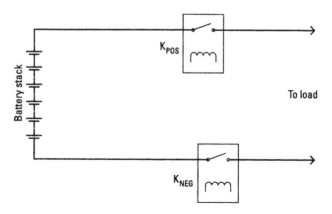

**Figure 7.7** Dual contactor layout.

## Control

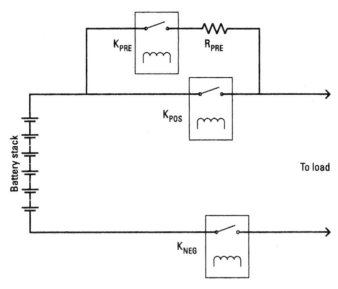

**Figure 7.8** Positive/negative/precharge contactor layout.

to determine if a conductive path exists between the two sides of the contactor. The other is to use the position of auxiliary contacts installed in the contactor which move synchronously with the main contacts but are designed to make a low-voltage, low current control circuit which can be monitored by the battery management system to determine if the contactor has operated correctly.

The high-voltage measurement method requires measurement of the high-voltage bus on both the battery and load side of the contactors (see Figure 7.9). If the system requires this measurement to be made for other purposes (which is often the case), then this does not impose a new requirement. This method indicates with certainty if a conductive path exists across the contactor, and therefore can reliably detect welded or opened contacts regardless of the contactor's condition. There are some challenges that will be discussed to ensure

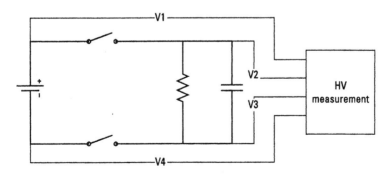

**Figure 7.9** Measurement topology for fault detection.

that reliable results are obtained with a capacitive load. Auxiliary contacts have the principal disadvantage of adding cost to the contactor and requiring an additional digital input to monitor their state. New failure modes are also created from situations where the measured state of the auxiliary contacts does not reliably indicate the state of the main contacts. This can happen if auxiliary contacts are burned due to excessive current or do not close due to oxidation (many auxiliary contacts are "wet" contacts requiring a minimum current when closed), or due to wiring faults or contactor damage.

In general, the auxiliary contact strategy requires additional inputs and wiring and leads to a wider variety of failure modes associated with improper operation of the auxiliary contacts. The high-voltage measurement method is a direct detection of the contactor circuit's current state and has a much higher fault tolerance.

Consider a two-contactor topology using a contactor at the negative and positive terminals of the battery. Four high-voltage measurement locations are required to provide contactor fault detection in this scenario, two for the battery voltage and two for the load voltage.

Contact welding usually occurs during contactor closing, but may occur while contactors are being held closed as well. Contactor welding tests should be performed at system startup (to ensure that a defective part was not installed during manufacturing or service) and after attempting to open the contactors.

Contactor welding faults can be intermittent because light "tack" welding can be released with mechanical vibration or shock which may occur during transport of the system for repair. Contacts that have been welded once will usually display a higher susceptibility to welding again, and it may be desirable to disable the battery system until the contactor is replaced and the fault condition is cleared rather than return the system to normal operation if the welded condition is no longer detected.

If a single welded contactor is detected on startup in systems with redundant contactors, there may not be an immediate safety issue, but it may be impossible to close contactors while maintaining soft-start functionality (in the case of the POS-NEG-PRE topology, if the positive contactor is shorted, for example). If a proper soft-start cannot be maintained, the load device may be severely damaged and a fire may result, and the battery should not be connected to the load under any circumstances. If a second contact welding failure occurs, the battery is now permanently connected to the load and the terminals of the battery are electrically live at all times. In most cases, battery operation with a single welded contactor should be inhibited until the failure is repaired. At a minimum, the need for service should be communicated to the system operator so that the contactor system is repaired quickly.

Failure of a contactor to close can be the result of a wiring harness fault, internal contactor coil open circuit, contact damage, or disconnection of high-voltage terminals. These failures can occur at any time regardless of whether the contactors are closed or not, or whether the battery management system is actively monitoring the system. As such, checking that contactors which have been commanded to close have actually done so should be done after the command to close is issued, as well as during operation to ensure that they remain closed. Combining this with coil-current measurement can identify the difference between a coil fault and a high-voltage contact fault.

A typical contactor sequence can proceed as follows:

- *Welded contactor detection:* Initially verify to ensure that $V_{12}$ is present (pack voltage) and $V_{34}$ (load voltage) is zero indicating the load capacitor is fully discharged. Measure $V_{32}$ and ensure zero reading (indicating positive and precharge contactors are not welded). Measure $V_{14}$ and ensure zero reading indicating negative contactor is not welded.

- *Contactor closing sequence:* Perform welded detection, if needed. Battery limits should be set to zero and/or load should be commanded to draw zero current. Close negative contactor and verify $V_{14}$ is equal to $V_{12}$ (indicating negative contactor closes correctly). Close precharge contactor and begin to monitor $V_{34}$, which should rise exponentially according to $V=$ ... If $V_{34}$ does not rise within the expected tolerances, or $V_{34}$ does not rise to within $V\Delta C_{max}$ of $V_{12}$, the closing sequence should be aborted. In some cases, if the load voltage rises immediately, the sequence should also be aborted as it is indicative of the proper load not being attached. When $V_{34} > V_{12} - V\Delta_{max}$, then the positive contactor can be closed. Precharge contactor can be opened shortly after allowing enough time for positive contactor to physically close.

# 8

# Battery Management System Functionality

## 8.1 Charging Strategies

In many large-format applications, the battery management system is at least partly responsible for controlling the battery charging. Control authority could vary from a very battery management-centric viewpoint in which the battery management system makes all decisions about charging and the battery charger electronics only perform the desired power conversion under command of the battery management system, to a charger-centric implementation in which the battery management system enables the charger which is then responsible for most of the decisions during the charging process.

The important decisions that the battery management system must make are the rate of charge to optimize utility of the system (faster charging allows the system to be discharged sooner), life of the batteries, overall efficiency, and other factors, and when to terminate charging.

### 8.1.1 CC/CV Charging Method

The CC/CV method (see Figure 8.1) is often discussed as the preferred method for charging lithium-ion batteries. Following this method, the batteries should be charged at constant current (recommendation depending on the particular cell, temperature, and other factors) until the end-of-charge voltage is reached. When this voltage is reached, the charging should switch to constant voltage, during which the current will gradually taper off as the cell charges. The charging is complete when the current tapers to a predefined level.

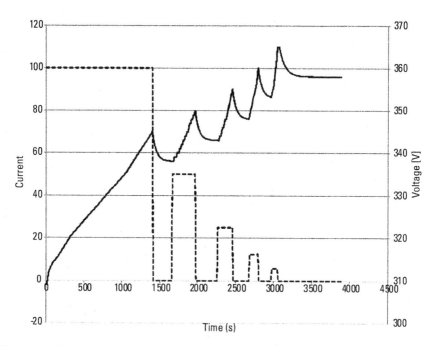

**Figure 8.2** Target voltage charging.

is charged at the maximum capability of the battery charger or the maximum recommended charge current for the cells (whichever is lower) until the highest cell voltage in the pack reaches the target voltage. The current is then scaled back and the target voltage increased and the process repeated. The target voltage should approach the CV voltage asymptotically (a "half the distance to the final CV voltage" strategy has been used with success) and the current will gradually taper off.

If the target voltage is very close to the final CV voltage, and the current is not reduced enough when the target voltage is reached, the cell voltage may exceed the CV voltage creating a mild overcharge condition. This can be prevented by cutting the current to nearly zero and implementing a pause and allowing the cell voltages to relax slightly before resuming charging.

### 8.1.3 Constant Current Method

For many simple battery systems, battery charging is performed at a fixed current. This is not as common with the introduction of lower-cost and higher-capability power electronics, but a variation on the target voltage method can be used to charge lithium-ion battery systems safely even if no current regulation is available. Figure 8.3 shows a constant current changing method.

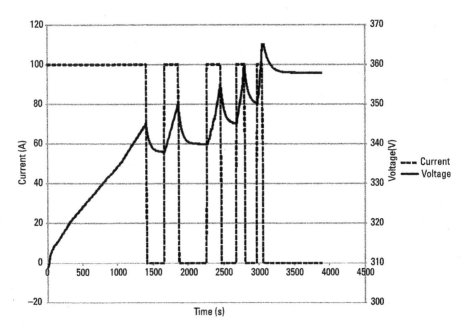

**Figure 8.3** Constant current charging.

The target voltage is established as above. When the target voltage is reached, the current is cut off and the cells allowed to relax until the maximum voltage falls below a restart threshold, at which point the current flows again. The target voltage is raised in the same manner as in the target voltage method.

Because the current cannot be reduced, the charging time for each current pulse will be successively shorter as the battery approaches full charge. The relaxation period will also increase in duration.

## 8.2 Thermal Management

In a large-format battery system, the battery often requires thermal management and the battery management system is expected to fulfill the control functions.

The inputs required for thermal management control include battery temperature measurement, which is required for basic safety and functionality, as well as possibly additional temperature measurements for inlet and outlet air or coolant. When pumps, valves, or fans are controlled, a feedback signal is often incorporated to verify that they are working as intended (pumps or fans usually employ a pulse width modulation-type tachometer signal with a frequency proportional to the rotational speed of the motor) to allow diagnosis of failures. Humidity measurement can be used to prevent excessive condensation inside the battery system if a dehumidification strategy can be employed.

Whenever battery heating is involved, multiple levels of safety are warranted to prevent the heating components from causing the batteries to enter thermal runaway. This is a very dangerous situation as the overheating will be widespread throughout the battery pack and may lead to multiple concurrent thermal events. Single-point relays or contactors can fail in the contacts-welded state leading to uninterruptible heating. Multiple transistors should also be used to control contactor drive circuits to prevent driver failure from leading to contactors inadvertently closing.

## 8.3 Operational Modes

Most battery management systems will implement one or more finite state machines responsible for controlling the operational states of the battery. These state machines will respond to external commands as well as to the detection of various conditions internal to the battery system.

A "low-power" or "sleep" mode will often exist. In this mode, the battery is disconnected from the load through the opening of the contactors, and the battery current is therefore necessarily zero. The system should minimize the consumption of energy both from the high voltage battery stack and control power supply. There is no need for monitoring of cell voltages and temperatures since the battery system is disconnected, and therefore all monitoring circuits and ICs should be in a high-impedance state. Communications buses are idle and in a low-power state. Microprocessors should be powered down and as many circuits should be deactivated as possible. An electric vehicle with the ignition switched off is a good example of where such a mode would be used.

A periodic wake-up from sleep mode, even if the battery system is not being activated, is useful for many reasons. As cell dynamics relax, a more accurate estimation of state of charge and cell balance is possible. Self-test functions can be performed on the battery management system and the battery cells can be checked for various defects. To implement this, a real-time clock or timer circuit is required in the hardware. Very low-power devices exist with an "alarm-clock" function that can be set from the main processor using a serial bus such as $I^2C$ (integrated circuit) or serial peripheral interface (SPI).

In some cases, cell balancing can be performed in this sleep mode. If this is the case, a periodic wake-up to ensure that the balancing is proceeding correctly. Many battery systems will spend a significant portion of their lifetime in this state.

An "idle" or "standby" state also usually exists. In this mode, the battery is still disconnected from the load, but the monitoring circuits are active. Cell voltages and temperatures are being measured, fault detection algorithms are operating, and state of charge, limits, and other state estimation algorithms will operate. This state allows the condition of the battery cells and entire system to be verified while the battery system is disconnected and charging and discharging are prevented. This state may be used on startup and shutdown to ensure that the system is safe before closing contactors. Cell balancing could be performed in this state if needed. Communications buses are typically active and the battery is exchanging information with the load and other devices on the network—this allows commands to be received and data such as fault status to be retrieved. High-voltage devices connected to the high-voltage bus should not expect a high voltage to be present.

If the battery performs a precharge or soft start, a special operational mode may be required for this. In this mode, other devices on the high voltage bus should expect the bus voltage to rise to the battery voltage, but must not consume any current from the bus to prevent a precharge failure. The contactor closing sequence would be triggered in this mode, initiated by an external command to connect the battery. The state would end in either successful completion of the contactor closing sequence, or in a fault which is detecting during the attempt.

There may be multiple online states where the battery is connected to a load or charging device. Many applications such as grid storage and hybrid-electric vehicles use the same network of devices to charge and discharge the battery and therefore there is no distinction made between a charging mode and a discharging mode. Other systems with a separate load and charger may use different modes for connecting each device (battery electric vehicles are an example).

An error state may exist in which the battery is idle with contactors open but does not respond to certain commands due to problems in the battery system. In this mode, the system can be interrogated for fault codes and diagnostic routines performed, but until a clear command is executed to leave the error state, the system cannot close contactors. Other possible implementations will allow connection but inhibit charging and/or discharging. It is important to avoid endless cycling between an error state and attempting to connect the battery; this can occur if certain battery faults clear automatically, triggering an attempt to return to the active state.

For battery systems with thermal management, a preconditioning cycle may be performed before the battery is brought online if the temperature of the battery needs to be adjusted to a certain target before the battery can be used. In this mode, voltage and temperature measurement are active, thermal management elements (fans, pumps, heaters/chillers) are active to change the battery temperature, but the contactors can be kept open to prevent current flow into a battery which is outside of the permissible operating temperature range. This obviously prevents energy from the battery from being used to power the heating or cooling and requires an external power source for this type of architecture. An example state diagram is shown in Figure 8.4.

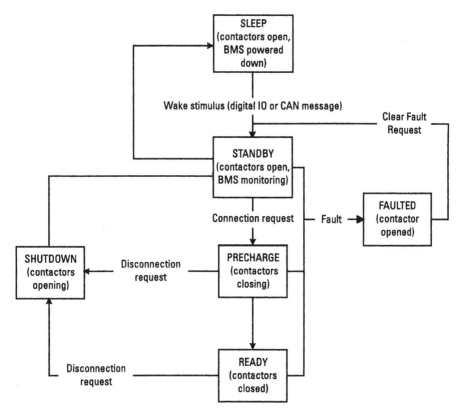

**Figure 8.4** Typical battery management system state transition diagram.

# 9
# High-Voltage Electronics Fundamentals

## 9.1 High-Voltage DC Hazards

A great deal of literature and institutional know-how exists about safety hazards for high-voltage ac systems found in nearly every commercial and industrial environment. With the advent of modern battery systems and power electronics, the existence of high-voltage dc systems is becoming increasingly common. Battery management system designers should be aware of important differences in safety aspects of ac and dc high-voltage systems.

Interruption of dc currents in high-voltage systems is more challenging than ac currents. Because the current does not go to zero with each half-cycle as with ac, electric arcs created are not self-extinguishing. Relays and contactors for dc are specifically labeled as such and generally include features such as hermetically sealed enclosures filled with hydrogen and magnetic arc suppression/blowout. The dc disconnects are specially designed to break the dc arc when opened. Never use a contactor, disconnect, or relay that is not rated for dc to interrupt dc.

Electrolytic electromigration, or "dendrite" growth, is the transport of conductive material across an insulating surface driven by an electric field. This phenomenon can cause unwanted shorts (as dendrites reach across insulating barriers and contact other conductive parts) or open circuits (as conductive material is removed from traces). The rate of growth is dependent on the applied voltage, and the growth of dendrites is significantly more pronounced using high-voltage dc. Dendritic growth (see Figure 9.1) requires surface moisture and ionic contamination to initiate the process. High-voltage electronic

**Figure 9.1** Dendrite growth.

PCB assemblies should therefore be manufactured, handled, and installed with maximum care to prevent excess moisture and contamination.

## 9.2 Safety of High-Voltage Electronics

There are important differences in implementation between "normal" embedded control systems (which generally operate with low-voltage control signals between 5-V and 28-V dc) and high-voltage devices such as battery management systems for large-format lithium batteries.

Most low-voltage embedded control systems have relatively low potential for electric arc hazards caused by dielectric breakdown or electric shock. High-voltage electronics must consider both of these potential risks. The potential for thermal events or fire exists with even low-voltage electronics and increases significantly as the maximum voltage encountered rises.

The most obvious concern associated with high-voltage electronics is that of dielectric breakdown between two conductive parts at a large potential difference. This can occur within the context of a single printed circuit board, between traces, component leads, connector pins or mounting features. Breakdown can also occur between PCBs, between PCBs and housings, between connector pins, or between any two conductive parts—not all of which are necessarily intended to carry current.

It is dangerous to simply apply the standard breakdown values for dielectric materials. Local field strengths around sharp edges may be much higher. Breakdown resistance is reduced with increasing humidity and contamination.

The most appropriate way to prevent breakdown hazards is to ensure that the product design process ensures that appropriate *creepage and clearance* distances are maintained.

Creepage is defined as the minimum path length between two conductive parts measured along the surface of a dielectric material. Creepage distances are required to prevent breakdown due to tracking, the creation of a conductive dendrite on the surface of an insulator. Figure 9.2 shows the difference between creepage and clearance distances between two conductive parts.

Creepage distances can be increased by notching or slotting the PCB substrate. The comparative tracking index (CTI) of a material is a relative indicator of the susceptibility of the material to electrical breakdown along the material's surface. The tests to establish the CTI of a given material measure the voltage required to cause tracking under standard test conditions. The CTI of the insulating material can affect the required creepage distance at a given working voltage.

The prevalence of tracking is increased with the following factors:

- High ambient humidity.
- Presence of contaminants or corrosive materials.

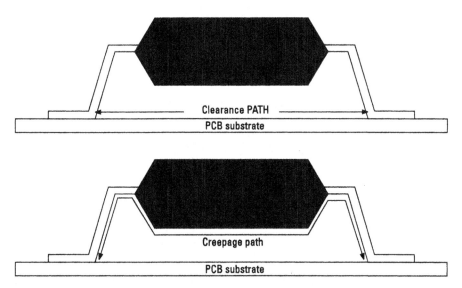

**Figure 9.2**  Creepage versus clearance distances.

- Low ambient pressure is a somewhat counterintuitive phenomenon that is a common condition with a high service altitude. Ground transportation may encounter altitudes up to 12,000 feet. Aviation may encounter altitudes of 50,000 feet or more. Aerospace applications may need to operate down to the vacuum of space.

Clearance is defined as the minimum path length between two conductive parts measured through air.

A number of standards exist that define the minimum creepage and clearance distances for a given working voltage and pollution degree. IEC 60950 and IEC 60664 are the most common of these standards.

The *pollution degree* defines the characteristics of environment in which the device operates. A summary of the pollution degrees defined by IEC60950 and IEC 60664 are next.

Different classes of insulation are defined by this standard. Basic insulation is the minimum level of insulation suggested to provide basic insulation against dielectric breakdown and electric shock. Double insulation, as its name implies, requires two independent barriers between high voltage and user-accessible parts. Reinforced isolation is a level of isolation provided by a single barrier which is equivalent in breakdown resistance to double isolation. A good guiding principle is that under normal operation and under single-point fault conditions, there should be no conductive path between high-voltage and any part which could be contacted by a user. Standards such as IEC 60950 provide clear guidance on how these standards should be met for grid-tied and other types of electrical equipment. The standards do not apply to all applications (automotive high-voltage battery systems are one exception) and no equivalent creepage and clearance standard has yet to be proposed. Therefore, the use of these standards in applications excluded from their scope should be accompanied by sound first-principles engineering and a great deal of testing.

IEC 60950 and IEC 60664 provides minimum creepage and clearance values for a given material class, working voltage, pollution degree, and insulation level. Methods for using this standard effectively are discussed next.

The use of coatings can be used to improve the resistance to breakdown. Despite the high breakdown strength of conformal coatings, there are a number of areas in which care must be taken when using coated PCBs with high voltage.

Coatings can be poorly or inconsistently applied with reduced effectiveness. Machine application of coatings is recommended for maximum uniformity. Inspection of coated boards can be performed under UV light to detect uncoated areas. The selection of the inspection frequency depends on the application and may vary from random spot checks to 100% inspection for the most safety critical components. Standards, such as IPC-A-610D, define acceptance levels for defects in coatings. Automated equipment for inspection of conformal coat can increase the speed and repeatability of the inspection process and check the coverage and thickness of coatings. Where coatings are relied upon to meet creepage and clearance requirements, an appropriate plan to ensure the coating quality and performance must be in place.

Component packages should be selected with creepage and clearance requirements in mind. The smallest available package may not offer adequate spacing. In some cases, even UL-listed semiconductor devices still require a double-check of lead/leg spacing.

In the vast majority of cases where the BMS is implemented on a PCB, most high-quality PCB layout software can perform design rule checking on the final layout artwork. This is an important step in aiding the verification of creepage and clearance distances. Some software packages are only capable of performing this analysis on the printed traces and pads, which is only a partial solution. Other packages or combinations of PCB layout software and three-dimensional modeling tools can provide a full three-dimensional analysis of creepage and clearance between PCB traces, component leads, connectors, mounting hardware, and enclosures. Demonstrating that this important contributor to product safety has been properly implemented is a best practice that should be followed for any high-voltage electronic system.

The following is a summary of best practices recommended for creepage and clearance safety on high-voltage circuit boards.

- Refer to IEC 60950, IEC 60664, or a similar standard for creepage and clearance requirements.
- Determine the appropriate pollution degree based on the installation in the application. A sealed enclosure may provide a lower pollution degree and therefore reduce the creepage and clearance distances.
- Select the desired insulation type.
- Determine the voltage levels and reference portals for individual elements of the system. Segregate areas of the circuit board according to voltage level and reference voltage.
- Use an automated method wherever possible to determine the minimum distance between parts for evaluating creepage and clearance.

## 9.3 Conductive Anodic Filaments

Conductive anodic filament (CAF) hazards are also a concern with high-voltage electronics. Conductive anodic filaments can form on printed circuit boards in humid conditions with high voltages between two plated through-hole features (vias or component pads) due to electromigration of copper ions from the PCB traces along the woven glass fiber of the PCB, producing a conductive filament which can cause an unintentional connection between two points at a large potential difference. To prevent CAF hazards from occurring, a number of practices should be followed in design and manufacturing of printed circuit boards exposed to high voltages.

The copper filament conducts due to the presence of copper ions in a moist medium. Unlike dendrites which are metallic and conduct electronically, the CAF conductivity is ionic.

Since the filaments form along the warp and fill directions, care can be taken to prevent aligning vias or through-holes with the glass fibers. Offsetting vias and holes will help avoid CAF formation if the glass fibers do not contact multiple vias at different potentials.

The quality of the laminate also influences susceptibility to CAF hazards. CAF formation requires separation of the glass and epoxy which produces a path for ingress of copper ions and growth of the CAF. Absorption of water by the laminate can increase the susceptibility to this phenomenon. The manufacturing specifications for PCBs for high-voltage applications should clearly specify laminate materials which are appropriate for this application. IPC-4101B specifies requirements for base materials for PCBs, including quality levels and test methods.

Unlike dendrites which grow on the surface of PCBs, CAFs are below the surface and cannot be readily seen. CAFs grow from an anode to a cathode, whereas dendrites grow from cathode to anode.

IPC provides a test procedure (described in IPC-TM-650) that should be performed on PCBs to be used in high-voltage applications.

## 9.4 Floating Measurements

There are important peculiarities about circuits designed to measure voltages that are not referenced to "earth" or "chassis" ground. In theory, the choice of any particular point in an isolated battery stack as a reference voltage against which to make cell measurements is as good as any other reference. However, in reality, many systems with isolated high-voltage components contain parasitic capacitances and conductances between isolated components and the earth/chassis ground.

These parasitic elements introduce a number of complications to the design, analysis, and testing of large-format battery management systems.

### 9.4.1 Y-Capacitance

The term Y-capacitance is used traditionally to describe capacitors placed intentionally between the ac input of a power supply and chassis ground (see Figure 9.3). In the context of battery systems, Y-capacitance refers to capacitance, either intentional or parasitic, which exists between the high-voltage system and earth ground.

While these capacitors are generally placed for reasons thought to be beneficial, they can create a number of problems. If they become too large, Y-capacitances pose a hazard to safety.

Consider the following isolated battery system and its isolated load, with Y-capacitance installed between the two terminals of the load and earth ground. Assume that a fault is introduced by a person making unintended contact with part of the high-voltage system and the earth ground, as shown in Figure 9.4. The resistance of the fault is $R_{fault}$. Assuming that the capacitor C2 is uncharged, in a transient analysis, it can be replaced with a short circuit. The resultant circuit shows that the fault resistance will be connected through the discharged capacitor across a high potential.

The energy that flows in this transfer is limited by the size of the Y-capacitance. Standards such as IEC 60479 detail the safety level that is associated with electric shock hazards of different magnitudes.

Y-capacitances also pose a similar safety hazard when the system contains switching power electronics that generate ac voltages that can flow across capacitors that block dc voltages.

## 9.5 HV Isolation

In many large-format systems, the dc battery voltage is not referenced to any earth or chassis ground. In the case of EV and HEV automotive battery systems, the high-voltage battery is not referenced to the 12-V chassis ground system at

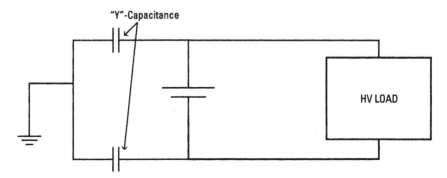

**Figure 9.3** Battery system with Y-capacitance.

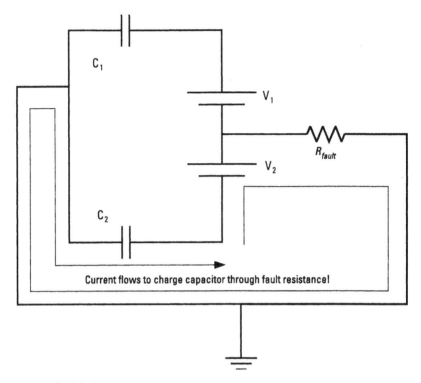

**Figure 9.4** Fault in the presence of Y-capacitance.

all. This provides an extra measure of safety in that a single point fault where the HV system is connected to earth ground does not create a large fault current.

As such, electronic systems that are connected to the battery cells will need to provide electrical isolation barriers between the battery cells and earth ground. Power and signals will need to be transferred across these isolation barriers without compromising their isolation.

On either side of an isolation barrier, the signals are referenced to different ground potentials that are not electrically connected. Because there is no fixed potential that can be used to measure the potential of these two systems, the potential difference between the two systems is arbitrary. It is often said that these two electrical systems are then "floating" with respect to one another, such that there may exist a significant potential difference between the reference (or "zero") potentials of the two systems.

Signals can be transmitted using different types of signal isolators. These can be separated into optical isolators (the signal is converted into light which is detected by another device on the opposite side of a transparent isolation barrier), capacitive isolators (a high-frequency signal is conducted across a small dielectric barrier that blocks dc), and magnetic isolators (electrical signals are converted to magnetic fields which cross a galvanic isolation barrier).

Digital isolators are suitable for transferring high-speed digital signals such as RS-232, RS-485, SPI, I²C, or CAN communications across an isolation barrier. Each isolation circuit is generally capable only of transmitting a signal in a single fixed direction, but devices are available with multiple channels in various combinations of directionality to suit different communications strategies.

High-speed digital isolators can create high radiated emissions if proper layout guidelines are not followed. Using digital isolators such as the ADuM™ family of devices requires capacitive coupling between the two sides of the isolator to defeat the parasitic dipole antenna created between both sides of the isolation barrier.

Low-speed signals are usually isolated using optical isolation. The most common optoisolator is a light-emitting diode (LED) coupled with a phototransistor in a common package. A dielectric insulating material is used between the two components. Because the switched side of the optoisolator uses a transistor device, the current on the switched side can flow in only one direction. This is suitable for isolating signals but does not provide the bidirectional power flow required for many power connections.

Optoisolators are available with different isolation ratings. There are usually two ratings available, the working voltage, which is the maximum voltage between the primary and secondary side during normal operation, and the dielectric withstand voltage, which is the maximum allowable applied voltage (usually limited in time, frequently to 1 minute) between the two sides without risk of dielectric breakdown occurring. The working voltage rating should be used for the maximum voltage encountered in normal device operation and the withstand rating should be used for the maximum fault or abnormal condition that must be withstood. These ratings are based on both the creepage and clearance distances as well as the materials used for the optoisolator package. The clearance distance is usually internal to the component, measured between the two sides of the device. The creepage distance will be measured across the outside of the package.

Optoisolated relays allow bidirectional current flow due to a complementary pair of transistors on the switching side. When the transistors are switched on, a bidirectional low-impedance connection exists, similar to a pair of contacts in a mechanical relay. Optoisolated relays are available in a variety of current ratings and voltage ratings. The relay will be rated for a maximum voltage blockage rating on the secondary side as well as a maximum voltage differential across the isolated barrier.

Optoisolators and optoisolated relays used in a switching application will have a specified turn-on and turn-off time that in many cases will not be equal.

All optically isolated components require careful analysis of the drive circuit. The LED must receive adequate drive current to ensure that the phototransistor fully turns on. Ideal optoisolators for switching digital signals would have infinite current transfer ratio (CTR), but in reality the switched current depends upon the drive current. As the device ages, the CTR falls and the drive current to achieve a given secondary current increases. LED degradation is worse at higher temperatures and higher operating current. Optoisolated devices must be designed to provide adequate performance over the entire life of the device. Some optocouplers have an input driver IC which accounts for degradation of LED output.

Optical isolators also introduce parasitic capacitance between the primary and secondary sides. This capacitance must be considered in the Y capacitance budget if a significant number of optical isolators are used.

In general, the selection of isolating components is a critical task in battery management system hardware design. Many components are rated to meet the requirements of standards established by various regulatory bodies which may be required depending on the application of the battery system. Isolator components are intended to keep hazardous voltages from reaching user-contactable surfaces and low-voltage circuits that are not equipped with adequate insulation and protection for high voltages. Ensure that the following information is available when selecting any component designed to cross an isolation barrier:

- Expected maximum working voltage across isolation barrier;
- Expected maximum breakdown/overload/test voltage across isolation barrier;
- Required standards to be met by the devices;
- Required creepage and clearance across isolation barrier.

## 9.6  ESD Suppression on Isolated Devices

Most electronic modules require testing to demonstrate they are capable of resisting electrostatic discharge (ESD), which can be created during use or during handling.

ESD consists of an effective high-voltage (typically between 4–25 kV) RC discharge into the circuit in question. If they are subjected to high voltages, even though the amount of energy in an ESD event is small, semiconductor devices, especially MOS transistors, can be permanently damaged. The voltage created during an ESD event, whether in service or during testing, is applied between the circuit in question and earth ground, and can be of either polarity.

The standard method of preventing damage is through the use of capacitors, diodes or transient voltage surge suppressors (TVS) devices mounted close to connector pins that provide a low-impedance path for the ESD event to be conducted to ground. The circuit is usually mounted in some type of protective enclosure which prevents ESD from reaching sensitive components through other paths which are not protected in this way.

If isolated circuits are included in the electronic device that is not referenced to ground in any way (see Figure 9.5), the energy from the ESD event will not be dissipated to ground. High voltages may be applied across isolation barriers, creating breakdowns, and circuits may easily be damaged.

It is necessary to provide a path for ESD to reach earth ground, but how can this be done when the goal is to isolate the battery stack from earth ground?

The use of high-voltage TVS devices offers one possible solution (see Figure 9.6). TVS devices can be placed between the high-voltage inputs and earth ground. The TVS must have a breakdown voltage greater than the voltage expected on the inputs to ensure that the TVS does not conduct under normal operation, but this voltage must be low enough that the downstream

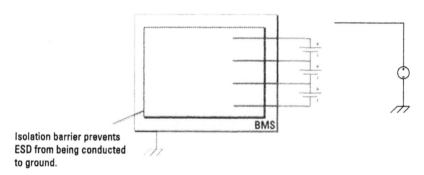

Figure 9.5  ESD path blocked by isolation barrier.

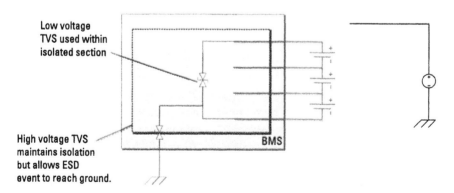

Figure 9.6  Possible ESD solutions.

components are properly protected. A cascading combination of TVS and capacitors may allow a reduction in the number of TVS required.

High-voltage capacitors are another option. These represent a significantly higher cost than low-voltage capacitors typically used for ESD suppression. These capacitors need to be installed between the high-voltage and low-voltage systems and therefore effectively span the isolation barrier, therefore increasing the system Y-capacitance. As the capacitance needed for ESD suppression is usually minimal, the added Y-capacitance is acceptable in many cases, but is worthy of consideration, especially if this strategy is used for every cell voltage measurement input.

## 9.7 Isolation Detection

On systems with isolated battery stacks, detection of an isolation fault or ground fault (an unintentional current path between the battery stack and components at or referenced to earth ground potential) is an important feature. Because the remainder of the system may not be energized if the battery is disconnected, it makes sense to place this functionality inside the battery system.

The method most commonly used to detect isolation faults is detailed next. It is the method used for static measurements of isolation in accordance with the Federal Motor Vehicle Safety Standard (FMVSS) 305 test performed on battery-powered vehicles.

Assume that a fault of resistance $R_{fault}$ exists between the battery system and earth ground (see Figure 9.7). This fault can occur at any potential of the battery system. Assume the general case where the fault occurs somewhere between the two terminals of the battery, effectively dividing the battery pack into two subpacks, and creating the following equivalent circuit.

The value of $R_{fault}$ can be calculated by making two measurements. For each measurement, a known resistor $R_{iso}$ is inserted between the positive or negative terminal of the battery and earth ground. The voltage across this resistor is measured.

This isolation measurement can be made periodically. Continuous isolation measurement is usually desirable to provide constant monitoring of this important safety parameter. However, it is important to understand the timing limitations of this circuit. Consider the same circuit, but include the presence of Y-capacitance between the high-voltage bus and earth ground.

When the measurement resistor $R_{iso}$ is inserted, it now forms an RC circuit using the Y-capacitance. The capacitor must fully charge before the measurement can be made. This establishes a minimum time required for making this measurement and therefore a maximum frequency at which it can be made.

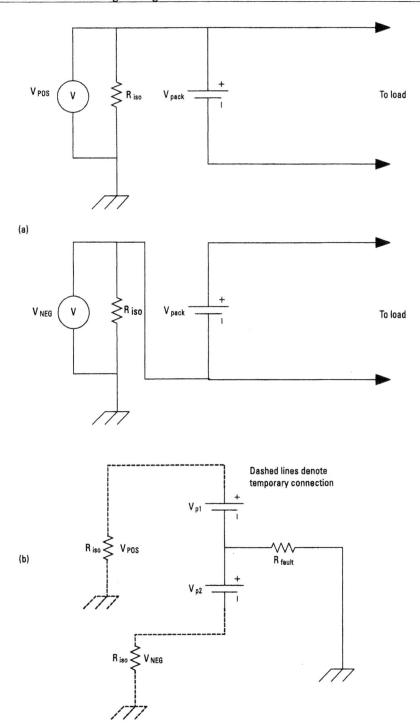

**Figure 9.7** Isolation resistance measurement.

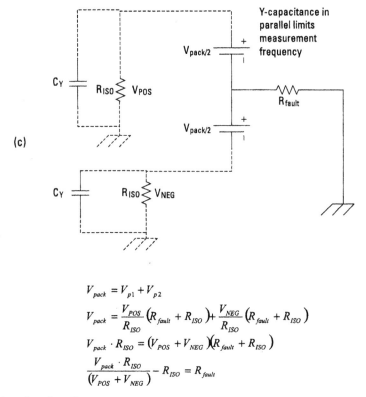

**Figure 9.7** (continued)

$$V_{pack} = V_{p1} + V_{p2}$$

$$V_{pack} = \frac{V_{POS}}{R_{ISO}}(R_{fault} + R_{ISO}) + \frac{V_{NEG}}{R_{ISO}}(R_{fault} + R_{ISO})$$

$$V_{pack} \cdot R_{ISO} = (V_{POS} + V_{NEG})(R_{fault} + R_{ISO})$$

$$\frac{V_{pack} \cdot R_{ISO}}{(V_{POS} + V_{NEG})} - R_{ISO} = R_{fault}$$

Isolation faults can occur due to the ingress of moisture into battery or electronics systems, due to the presence of foreign objects or materials (conductive debris) that bridge gaps between conductive parts, due to the formation of conductive bridges (see Sections 8.1 and 8.3) on circuit boards, faults internal to isolating components, insulation breakdown, and other hazards. Isolation faults are hazardous to those working on battery systems, and multiple distributed isolation faults at different potentials can create short-circuit hazards. Therefore, many applications specify that operation with compromised isolation resistance is prohibited, or must be accompanied by a warning to users. The battery management system itself is a place where high voltages and systems at different reference potentials are brought into close contact and faults may occur.

Isolation detection can be performed with the battery connected and disconnected to provide fault localization. If the isolation fault is external to the battery, opening the battery contactors will bring the system into a safe state. If the isolation fault has occurred inside the battery, even with contactors open, the possibility for hazardous conditions exists.

# 10

# Communications

## 10.1 Overview

Many battery management systems use a serial communications link to communicate with the load device as opposed to multiple discrete and/or analog signals.

Battery management systems will also use a serial link to communicate between master and slave devices within the battery pack.

The advantages of these communications protocols are well known (smaller number of wires required, high data transfer rates over relatively inexpensive connection, robustness against message loss) and the risks associated are also equally well documented (potential of data loss or corruption, higher susceptibility to EMI).

An overview of the various network technologies commonly used, as well as the types of information that are usually transferred via a communications bus, follows next.

## 10.2 Network Technologies

The Open Systems Interconnection (OSI) model is often used to partition the functionality in a communications system into various abstraction layers. Each abstraction layer implements a specific subset of the total connection, from physical signaling (voltage levels and timings), addressing, flow control, data representations, and finally application. There are a total of seven layers in the OSI model.

Various communications technologies exist for each of the layers. Some of the most common will be described here.

### 10.2.1 I²C/SPI

These protocols have been developed for communication between ICs within the context of a single printed circuit board. Although they are certainly acceptable for this purpose and may likely be robust enough to handle communication over short distances, I²C and SPI are not recommended for communication between circuit boards.

It is worthwhile to discuss the "level-shifted" communications buses offered by many IC manufacturers for communication between battery monitoring ICs. These buses may be level-shifted versions of well-known buses such as SPI or I²C, or they may be proprietary in design. Many of these manufacturers advertise that this communications method is suitable for connecting multiple ICs across significant distances of a meter or more to implement a distributed BMS architecture. If this approach is to be followed, in lieu of a more traditional intermodule communications technology such as CAN, RS-232/485, and so forth, it should be done with care and ensure that an appropriate level of validation and analysis is conducted.

A more robust and tried-and-true distributed architecture would use the level-shifted bus between a smaller number of ICs on individual PCBs/modules and the use of CAN or another communications bus between the slave modules and the master device. This will require the use of isolation components. Possible architectures are shown in Figures 10.1 and 10.2.

### 10.2.2 RS-232 and RS-485

RS-232 lacks differential signaling and was developed for point-to-point communication as opposed to a bus or star topology. Initially common for computer peripheral interconnection, this bus is seldom used in industrial, automotive

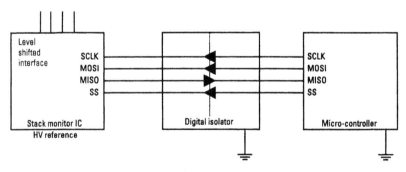

**Figure 10.1** Level-shifted SPI to isolated SPI.

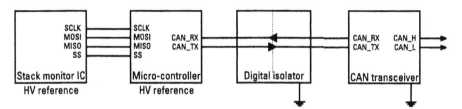

Figure 10.2  Level-shifted SPI to isolated CAN.

or commercial applications. The maximum cable length is limited to 15 meters (50 feet) unless special low-capacitance cable is used.

Many microcontrollers include a UART to support RS-232 and many resources exist for communicating between embedded controllers and PCs for development and debugging. Expensive hardware is not required, although most PCs have removed the native RS-232 port in favor of USB and other newer connection technologies. Still, RS-232 remains a popular technique for communications.

A dedicated wire is required for both the transmit and receive directions. RS-232 uses single-ended signaling, meaning that the information is contained in the difference in potential between the communications line and ground. This makes RS-232 unable to reject voltages induced on the signal line and susceptible to ground offsets (different in ground potential between receiver and transmitter). To overcome these challenges, high voltages are used (12V). In many embedded systems operating at 5V or lower, these voltages are generated by a transceiver such as the MAX232 using a charge-pump converter to produce 12-V output.

RS-232 lacks the low-level error checking present with CAN, and the single-ended signaling has increased potential for noise susceptibility. RS-232 should always be combined with additional services in a higher layer to ensure that errors in transmission are detected.

Flow control schemes such as software flow control using XON/XOFF or hardware flow control using RTS/CTS handshaking are often used. Higher speeds of devices mean that buffer overflows are less common, but flow control may be required for slower operations such as flash programming.

RS-232 implements the OSI physical layer. RS-485 uses identical timing parameters but with differential signaling as compared to RS-232. RS-485 finds common application as the physical layer used for Modbus communications, often used in commercial and industrial system controls. RS-485 still plays a significant role in industrial applications using the Modbus protocol stack and would be useful for battery systems that must interface with commercial and industrial electrical equipment. Modbus over RS-485 is common for many battery systems in the grid-tied storage sector. It has higher noise immunity and longer allowable cable lengths than RS-232. RS-485 allows the construction of linear networks (star or ring networks are not recommended with RS-485) as opposed to RS-232 networks, which are generally point to point. RS-485 networks are usually half-duplex, meaning that only one device is allowed to transmit at a time, usually requiring a single-master configuration.

### 10.2.3 Local Interconnect Network

Local interconnect network (LIN) was developed to provide an inexpensive serial communications protocol in automotive applications in which CAN was judged to be too expensive for all components for which OEMs wanted to add networking capability. LIN requires a master/slave topology where the slaves are often ASICs which do not require a microcontroller core or software. Data rates are relatively limited at 19.2 kbit/s but guaranteed latency times and other robustness features make it a possible choice for BMS master/slave communications.

LIN has been a common choice for simple, low-bandwidth sensor integration. This could have useful battery management system-related applications for integrating current and temperature sensors if a direct analog interface is not convenient. Sensors operating on LIN are becoming available for this purpose.

### 10.2.4 CAN

Controller area network (CAN), developed in the 1980s by Bosch, has found widespread adoption in the automotive industry and is also used in industrial applications as well, specifically when used with the CANopen protocol specifying higher layers of the network model.

In the OSI model, CAN specifies the physical and data link layers. The additional CANopen stack implements the network through application layers. CAN provides for high-speed communication allowing near-real time performance for many types of signals, and offers high robustness to EMI.

When used in control systems or automotive applications, CAN networks are usually referenced to earth ground. CAN nodes should be capable of operating with a limited ground offset voltage between different nodes on the bus.

CAN uses the message ID to specify the relative priority of each message being transmitted on the network. This can be used to resolve conflicts between interfering messages if bus loads are high. Vital communication links such as power modes and safety interlocks should use a high-priority message to ensure that they are not interrupted by lower-priority information.

CAN buses should typically be laid out in a linear fashion with a terminator at each end. If the battery management system lies at the end of a bus, it may need to implement the termination scheme. A split terminator with a central capacitor to ground allows a low-impedance path for common mode noise to ground.

Each node on the CAN bus is capable of transmitting messages and messages are received physically by all nodes. Most microprocessors are capable of using CAN filtering to use one or more bit masks to remove messages based on the message ID. In most networks, messages are sent continuously by all nodes at a rate dependent on the rate of change of the quantities represented in the message.

Many microcontrollers support CAN using one or more CAN ports; however, transceivers are required to adapt between the physical layer used at the microcontroller level and the physical layer used between electronic modules. The receipt and transmission of CAN messages by the microcontroller is often performed by hardware, low-level firmware, and interrupts, making exact time-stamping of message receipt and deterministic timing for sending difficult.

Automotive systems use sophisticated network management to allow devices to enter a power-down state and will require a wake-on CAN functionality, where the receipt of a CAN message will cause the controller to power up. Specialized transceivers are available with this functionality but the use of a wake-on-network strategy (using CAN or any other communications protocol) requires a system-wide coordinated wake/sleep strategy (often referred to as *network management*) to prevent parasitic power consumption.

The ISO 11898 standard specifies the requirements for the most typical CAN networks in automotive applications.

### 10.2.5 Ethernet and TCP/IP

Ethernet is the most common protocol specifying both the physical layer and the data link layer, used for residential and commercial computer networks and is beginning to see increased usage in industrial control networks as well. Ethernet has grown into a good choice for bandwidth-intensive but nonsafety critical communications.

Ethernet implements the data link layer, and is usually coupled with 10/100/1000BASE-T as the physical layer. Higher layers in computer networking typically use Transmission Control Protocol (TCP) and Internet Protocol (IP) for transport and network layers, and have migrated to industrial Ethernet as well and will have some relevance for industrial battery systems.

TCP was not initially intended for use in real-time applications due to the requirement for acknowledgements and the priority of accuracy (packets must arrive without error and in the order in which they are sent) over speed. User Diagram Protocol (UDP) is preferred when real-time performance is preferred. In battery systems in which responses are required in tens to hundreds of milliseconds, a real-time communications topology is certainly preferred.

### 10.2.6 Modbus

Modbus is an application-layer messaging protocol that describes communications between two devices. The lower levels can be implemented multiple ways, the most common being RS-485 and TCP/IP/Ethernet.

Modbus is used in the control and network of many industrial and commercial products. Modbus allows for a limited number of operations to be performed over the network. Information in Modbus-linked systems is transferred using Modbus registers.

Modbus requires a master-slave implementation as it uses a request/reply structure for all messages. Transactions are initiated by the device known as the *client* and the responding device is the *server*.

### 10.2.7 FlexRay

FlexRay is a relatively new protocol developed exclusively for automotive applications. It was designed to allow increased speeds and overcome a number of the limitations that are associated with CAN, namely, lack of deterministic timing, redundancy, fault tolerance, and time-triggered behavior. The higher data rates have somewhat limited usefulness for battery applications but the other improvements contribute greatly to robustness and bring a number of advantages for safety-related systems like batteries. At present, the use of FlexRay has not expanded beyond the automotive industry. The requirements of battery management do not require at this time require the advanced features of FlexRay.

## 10.3 Network Design

The network design must incorporate the transfer of data between the battery management system and the load device, and potentially a number of other systems as well, in addition to the communications between master and slave devices for distributed architectures.

The selection of one particular communications method (or multiple, as many battery management systems will include multiple communications techniques) will impact hardware selection (usually microprocessor support for the required communications ports are preferred to external implementations) as well as other decisions such as clock speed and accuracy required by various bus types. The network layout and number of nodes both internal and external to the battery management system needs to be considered. These factors will impact total system cost as well. In general, the transmission of battery data within a battery management system, and between the battery management system and other devices should not require high data bandwidths as compared with even modest networks for computer applications, for example. However, the reliability of these networks is critical and battery power applications require network technologies that are not prone to message corruption or loss under the expected conditions of their environment as the consequences of network failure are much more severe than failures in computer or telecommunications networks. For critical signals such as major fault status, redundant signaling using a secondary, more simplistic method may be advisable. Network latencies will also limit reaction times to commands provided over the network; for time-critical inputs (such as the vehicle crash condition discussed earlier), these latencies must be understood and analysis must show that the timing requirements are met by the chosen network topology.

As discussed previously, many battery system parameters do not need to be communicated externally and are only required to be sent between master and slave devices. Information such as cell voltages and temperatures is normally processed by the battery management system to generate externally useful information. The transmission of this data should be reserved to internal data buses if possible, since with a large number of cells, this will constitute a significant amount of data. It is useful to make cell voltages and other individual measured parameters available upon request for special diagnostics. To facilitate the correct routing of information, multiple communications ports or buses may be implemented. They may or may not necessarily use the same physical layer and/or communications protocol. During the development and testing of the battery system it is useful to provide a private bus containing significant amounts of diagnostic data which may be used in troubleshooting the system.

Communications between the master and slave devices include the cell voltage measurements, temperature measurements and cell balancing commands. Cell measurements are generally reported with approximately 1-mV precision, usually equivalent to the ADC precision in the measuring device, requiring 12 to 14 bits per cell voltage reading. Conversion to physical units is not necessary before data transmission. It is possible to use nonlinear compression (equivalent to companding) to compress the expected range of 0% to 100% SOC into fewer bits (voltages outside of the normal operating range do not necessarily need to be reported with the same precision). To ensure synchronization, cell measurements will often need a timestamp transmitted along with the voltage data to indicate to the master device when the measurements occurred. Temperature measurements are often made over a temperature range of -40°C to 80°C with a precision of 0.5°C–1°C, requiring 7 to 8 bits each. Again, a nonlinear compression scheme can be used for extreme temperatures or special codes can be used to represent off-scale readings (the master device needs to be aware of the difference between high but seemingly correct readings and those that indicate a hardware problem). Cell balancing commands may be communicated in different ways; the communications requirements for this depend upon the complexity of the balancing circuitry. For simple dissipative balancing systems, the master device can simply broadcast a single bit for each cell, which expresses the desired state of the balancing switch in real time. The update rate for cell balancing does not need to be more than once every few seconds. Alternatively, the balancing requirements can be communicated less frequently by sending the amount of charge to be removed from each cell and requiring the slave devices to count down the charge dissipated. This will work better for systems that operate continuously and are not subject to shutdown.

The network must be capable of transferring enough data between master and slave devices to ensure that the cell voltages and temperatures can be transported at full precision, at an update rate equal to their measurement rate.

For a battery pack consisting of 100 cells, each of which is measured with 12-bit precision every 40 ms, the minimum data requirements are: $100 \times 12 \times 1/0.040 = 30$ kbit/s.

All communications buses will have additional overhead required for message headers, start and stop bits, and checksums. These may double the data requirements. Timestamps and temperature measurements will have additional data requirements and sample rates may be faster. The example above could require up to 60 kbit/s. This is a high data rate for RS-232 or RS-485 over medium to long distances, but very feasible for CAN, as an example. This type of basic calculation can help determine minimum required data speeds and therefore help select a data bus.

Internal to the battery pack, because the load device for many types of systems consists of switching power electronics, bus bars may have currents containing conducted emissions starting in the low kilohertz range, with harmonics reaching up to hundreds of kilohertz. These frequencies are relatively low, but the field strengths may be very high due to high battery currents. These frequencies may interfere with certain data rates and single-ended (i.e., RS-232) buses are especially vulnerable. Internal battery pack communications harnesses should use twisted pair as an inexpensive method of reducing radiated susceptibility and avoid long coaxial runs of busbars or power cables and communications wires.

Communications with the load will nearly always include the battery charge and discharge limits expressed in terms of current or power. These will need to be updated frequently, especially in battery systems with high discharge rates that are used near the edge of their safe operating area, to prevent a limit violation. Update rates from 50 ms to 100 ms are common in many applications, with slower rates more feasible for more conservative battery usage or lower charge/discharge rates.

In all cases an expedient method of signaling to the load that the battery will be imminently disconnected and to command the battery system to disconnect immediately should be present. System safety will often rely on these two methods and the communications network should ensure that these messages are always responded to promptly. Examples include disconnecting the battery from an electric vehicle during the event of a crash or the battery management system notifying the load that the contactors will be opened due to a persistent safety fault that has not resolved itself.

For the following types of information exchange, the subsequent concerns for communications apply:

- *Cell voltage measurements:* Bandwidth (due to high number of cells, high precision and relatively high sampling rate), latency (rapid responses are usually required to overvoltage and undervoltage in the 10–100-ms range depending on application), network reliability, possibility of coordinating voltage, and current measurements in time through deterministic communication protocol, external synchronization signal, or master clock.

- *Cell temperature measurements:* Limited in bandwidth but requires high network reliability to ensure temperature measurements are not lost.

- *Current measurements:* Low in bandwidth due to small number of measurements but requires low latency and careful determinism. Often an analog interface makes sense for current sensors, but serial communications can be used as well. Consider discrete synchronization signal.

- *Redundant measurements/overvoltage or undervoltage/temperature signals:* These measurements do not require a large amount of bandwidth, but reliability is paramount when maintaining system safety.

If the base layer does not include a checksum or CRC arrangement (included with CAN and FlexRay but not with RS-232 or RS-485, for example), message integrity should be verified at the application layer to prevent acting on erroneous information.

Furthermore, for safety critical signals, of which there can be many in a battery management system, additional levels of integrity checks are often warranted to prevent against the following types of failure modes:

- *Prevention against "babbling idiot" failure modes:* To ensure that an appropriate action is taken if a module begins to transmit well-formatted but incorrect data (for example, cell voltages are reported in the correct format at the correct time, but due to an error in the part of the application, the same message information is repeated without updating the voltage values, despite the fact that the real cell voltages are changing), strategies can be implemented to ensure that "stale" data is not reported and acted upon as new. One simple scheme is a "rolling counter" in which a counter value is attached to each message. This counter increments in a predictable way with each new message. If the counter fails to increment several times, an assumption can be made that the measuring device is acting incorrectly. This approach requires special techniques to deal with the possibility of lost messages or in networks in which messages are not guaranteed to be received in the same order in which they are transmitted (possible with certain types of buffering and retry strategies). Other approaches exist with more robustness.

- *Checks on message format, data length, minimum and maximum values:* Messages that do not arrive in the correct format or appear to contain extra or missing data may be unreliable; consider discarding the data and bringing the system to a safe state. Especially in modern systems, which are developed with a high level of abstraction between low-level software drivers and model-based code blocks, values passed on communication networks often use physical units to represent quantities that may differ from one network node to the next. Particularly in development, it is important to ensure that critical data is checked against minimum and maximum expected values to prevent dangerous conditions. "Bit-stuffing" errors, in which bit fields are inserted or removed in a message of fixed structure, usually create erroneous data that can be easily checked.

- *Debounce or filtering may be useful on signals that can cause fast changes in state (connection and disconnection of the battery):* Ensuring that the same signal with the same instruction is received several times consecutively will minimize the risk of an inadvertent state transition.

Dealing with lost messages is an important design consideration. Many networks do not guarantee that messages will be retried until received, and some do not guarantee messages will arrive in the same order they are transmitted. Implementing safety-critical commands to and from the battery system should not be done using single messages if the network does not provide assurance that the message will be retried until received. The concept of most CAN networks is a more robust approach, where all messages contain quasi-real-time data, and are transmitted continuously. Failure mode analysis for important state transitions should consider the possibility of lost or unreadable messages.

A useful reference design for the battery connect request signal is as follows. This could be interpreted as the ignition-switch signal in an electric vehicle or the power-on request for an energy storage system.

The battery management system receives the primary instructions for the desired state of the battery system via the serial communications bus, as well as a secondary state signal using another means. The most basic implementation could be a single digital signal but this offers no way to detect a lost connection. A better method would be a signal with multiple analog windows, although this can be prone to EMI in certain cases and requires careful analysis to ensure that the signal is properly interpreted under all conditions, or a PWM signal with differing frequencies or duty cycles for different state requests. The secondary signal can be generated using a redundant circuit to verify that connection of the battery system is both desirable and safe, and could be routed in a physically independent path to reduce the likelihood of a common interconnect failure from severing both signal paths.

If complete loss of communications occurs, it may be necessary to either bring the system immediately to a safe state or to adopt a limited-operation strategy, perhaps temporarily, and wait to see if communications are reestablished. In some cases, a complete loss of communications is interpreted as a dangerous failure (automobile crashes often result in loss of network connections) and the battery is disconnected if such a failure persists for more than a few hundred milliseconds. In other situations, especially if a redundant signal is incorporated, the battery could adopt a limited operation strategy in which at least one piece of information exists that confirms the desired state of the battery.

11

# Battery Models

## 11.1 Overview

The complicated electrochemical and physical processes occurring in a battery cell lead to a set of externally observable behaviors. These behaviors exist as the relationships between voltage and current (as well as temperature, to some extent). As the primary purpose of the lithium-ion cell is an energy storage device in an electrical circuit, it is generally helpful to represent these behaviors as an equivalent electric circuit.[1] Other types of models may be used based on the fundamental physical and electrochemical processes that occur inside the battery.

A fundamental assumption in equivalent circuit modeling is that there is an observable state variable (or set of observable state variables) that is a function of state of charge. In most cases, the relationship between fully relaxed open-circuit voltage and state of charge is the one that is used. The assumption is often made this relationship is invariant with factors such as temperature, age and cycle life.

The term *observability* is used in the control theory context. Classically, an observable system is one in which the system's internal states can be fully determined in finite time (not necessarily immediately) from only the inputs and outputs of the system.

As the equations and models discussed in this section are often implemented in discrete-time systems, a discrete-time version of many of the concepts is

---

1. The same method is often used for other complicated devices in electric circuits. A circuit designer seldom thinks of a transistor in terms of the semiconductor physics at work; it is more useful to use a model that roughly represents the electrical behavior of the device.

presented. For concepts which only apply in discrete-time, the continuous-time version will be omitted.

In a battery management system implementation, the purpose of battery modeling is to create a method of transforming easily measured quantities (current, voltage, and temperature) into an accurate representation of internal states which are not easily measured, such as states of charge. A number of techniques are used for state of charge determination with other battery types, such as simple voltage look-up tables for simple applications and load profiles. It is inappropriate in large-format lithium-ion systems to perform state of charge computations this way, leading to the necessity for more advanced battery models.

Assume that the open-circuit voltage is a function of state of charge defined as $V_{OC}$ (SOC). This relationship is monotonic and defined such that a unique SOC maps to a unique $V_{OC}$. The relationship need not be linear, exponential, or even conveniently approximated by a closed-form mathematical relationship. This implies that if $V_{OC}$ is known, then SOC is also known. Furthermore, if the error in $V_{OC}$ can be bounded, the error is SOC is also bounded.

The purpose of equivalent-circuit battery modeling is to calculate the dynamic overpotentials (difference between the open-circuit voltage and the real, measured terminal voltage). If the overpotential is equal to $V_\phi$, then $V_{OC} = V_t - V_\phi$, allowing the SOC to be calculated from the measured voltage and the calculated overpotentials. Physics-based models may not make use of an empirical SOC-OCV relationship.

## 11.2 Thévenin Equivalent Circuit

In the most simple of analyses, the battery can be modeled as an ideal voltage source. An ideal voltage source is a two terminal device that has no internal impedance and has a fixed voltage across the two terminals, regardless of the current flowing, or any other parameters.

In many large format applications, the battery is used at a high discharge or charge rate, such that the assumption of an ideal voltage source is no longer valid. When operating a battery system at high currents, the first nonideal effect to consider is that of the cell's *internal dc resistance.*

The Thévenin equivalent circuit can be used as a simple battery model. An ideal voltage source, with a voltage equal to the battery's open-circuit voltage, is placed in series with a resistance, equal to the battery's dc resistance.

As such, the terminal voltage of this equivalent circuit is no longer constant; it will now depend on the battery current.

In reality, it may be the case that a simple dc resistance may not model all of the current-voltage relationships that are observed (see Figure 11.1).

# Battery Models

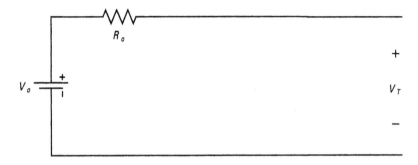

**Figure 11.1** Battery model with series resistance and corresponding voltage/current relationship.

- The resistance is likely to be a function of temperature and state of charge, and is likely to change as the battery ages.
- The resistance may be different during charging and discharging (see Figure 11.2).

To account for these factors, a pair of ideal diodes may be introduced to the equivalent circuit, and the resistance may be replaced with two variable resistances, which are a function of temperature, state of charge, and age.

Additionally, many large-format battery systems are operated over a wide range of state of charge. Therefore the effect of state of charge on the electrical behavior must be considered.

The equivalent circuit has therefore become "state dependent," that is, it has a number of internal states that influence its behavior; the system exhibits a "memory," which means the response cannot be predicted only from the system's inputs.

The ideal voltage source can be therefore replaced with a variable voltage source. The value of the voltage source is equal to the open-circuit voltage of the

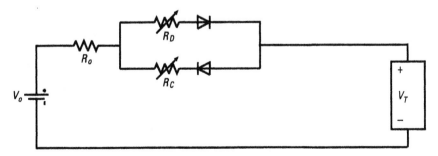

**Figure 11.2** Battery model with charge and discharge resistances.

battery at a given state of charge. The relationship between state of charge and open-circuit voltage is not necessarily linear.

All electrochemical cells exhibit a slight temperature dependence that can be predicted by the Nernst equation. The open-circuit voltage should change slightly with increasing temperature. In most applications, over the temperature range of interest, the significance of this effect is small, but for chemistries with a very flat SOC-OCV profile, incorporating this predictable effect can add accuracy.

Therefore, this version of the battery equivalent circuit looks like this:

- $V_0(SOC, T)$ is the open-circuit voltage, which changes with SOC (and possibly temperature) (see Figure 11.3).
- $R_c(SOC, T)$ is the charge resistance and $R_d(SOC, T)$ is the discharge resistance, both of which are functions of $SOC$ and temperature.

The following differential equation governs the relationship of state of charge and current.

$$\frac{dSOC}{dt} = \frac{1}{C}I$$

where $C$ is the capacity of the battery. It is clear that the rate of change of SOC is equal to the current divided by the battery capacity.

The relationship between terminal voltage, current, and SOC is given by:

$$V_T = V_{OC}(SOC,T) + IR(SOC,T)$$
$$V_T = V_{OC}(SOC,T) + I_C R_C(SOC,T) + I_D R_D(SOC,T)$$

where $I_D = I$ if $I < 0$ and 0 otherwise, and $I_C = I$ if $I > 0$ and 0 otherwise.

This model accounts for the changing open-circuit voltage of the battery as the SOC changes, as well as the ohmic resistance during charge and discharge

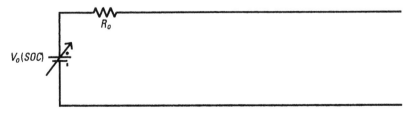

**Figure 11.3** Battery representation with variable voltage source.

currents. While this model may be appropriate for some applications, it fails to capture the dynamic response of the battery. A comparison of actual battery responses and those predicted by this type of model are shown in Figure 11.4.

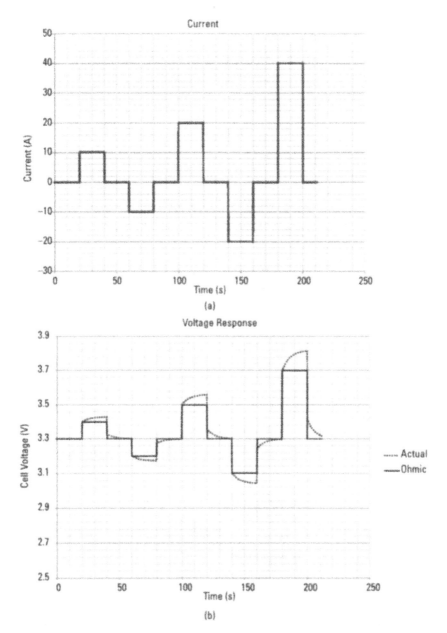

**Figure 11.4** Comparison of actual battery response and ohmic resistance model results.

Ohmic resistances of the electrolyte, electrode materials, current collectors, tabs, and terminals are described by a simple resistor and are well accounted for in this model.

The following list covers some of the effects which are not accounted for in this model:

- *Polarization or charge transfer resistance:* The rate of electrochemical reactions (in this case, the intercalation of lithium ions into the anode and cathode materials) is limited and proceeds at a rate depending on the applied voltage. The Butler-Volmer equation governs this effect. This effect can be modeled with a resistor known as the *charge transfer resistance*. The charge transfer resistance increases at low temperature.
- *Double layer capacitance:* The charge transfer resistance causes the accumulation of charge carriers at the surface of the electrode, which creates a capacitor-like effect due to the charge separation across a short distance.
- *Diffusion:* Concentration gradients exist in both electrodes and electrolytes leading to overpotentials.

To account for the electrical behavior caused by these phenomena, the *Randles cell* is commonly used (see Figure 11.5). The Randles cell consists of a single resistor in series with a parallel combination of a resistor and a capacitor. The resistance is often referred to as the charge transfer resistance $R_{CT}$, and the capacitance is referred to as the double layer capacitance, $C_{DL}$. It bears repeating that no real resistor or capacitor exist in the cells. Additionally, this model only approximates the dynamic electrical behavior of the cell. There are significantly more complicated phenomena which occur inside the battery that the Randles element only approximates. For some batteries and some applications, these techniques may not provide adequate performance. However, they are widely used and satisfactory performance can often be achieved using them.

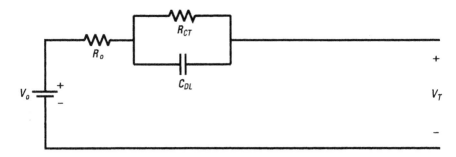

**Figure 11.5**  Randles cell equivalent circuit.

The response of a real lithium-ion battery to the application of an alternating profile, and the predicted response by a single-element Randles model with the given parameters are shown next. The model begins to approximate the dynamics seen in steady-state charging and discharging. Longer current pulses show increased divergence of the predicted and measured voltage.

To account for multiple time-dependent dynamic reactions occurring in the cells, the use of one or more RC elements in series can be used to improve the accuracy of the equivalent circuit. The time constant of the elements is different. Some cells can show dynamic behavior that takes several hours to fully develop. The $R$ and $C$ parameters are certainly temperature and state-of-charge dependent and they should not be assumed to be invariant over the life of the cell. If an online parameter estimation method is not available to dynamically update all of these parameters in real time, lifetime experiments can be used to determine which parameters are the most likely to require adaptive estimation.

Each RC element requires a single state variable to represent its state. This is usually the voltage across the capacitor, however alternatively the charge on the capacitor could be used. Typically the values of the resistor and capacitor may be nonsensical when compared to actual electrical circuit values with resistances much less than 1 ohm and capacitances in the thousands of farads.

The dynamic effects that are modeled by the RC elements are only approximated. The multiple-element RC circuit is a useful tool for establishing an electrical equivalent circuit that models the dynamic time-dependent performance of the battery. The circuit is a linear time-invariant system that has a number of advantages for simplicity of implementation and analysis, and is well understood by systems, electrical, and electronics engineers. For some batteries and applications, even multiple RC elements may not be capable of accurately modeling the battery's response, and other modeling techniques are needed.

## 11.3 Hysteresis

The presence of hysteretic behavior must be accounted for in a number of cell models. The presence of hysteresis makes the OCV-SOC relationship a path function and not merely a state function. A classic experiment is to take two identical cells, one fully charged, and the other fully discharged, and to approach equal SOCs from the opposite directions. In some cases, the measured voltages of these two cells will not be equal, even if the overpotentials are allowed to fully relax. The one-to-one relationship between SOC and OCV no longer holds. A method for modeling the hysteresis is therefore needed to accurately determine SOC if the battery in question has significant hysteresis.

A true relationship between SOC and open-circuit voltage is therefore impossible to achieve as changing the SOC necessarily requires current to flow.

A "neutral" relationship can be approximated by performing charge and discharge cycles at very low (and equal) rates, allowing for long relaxation times, and generating two curves. A curve drawn midway between the charge and discharge curves can approximate the zero-hysteresis voltage for a given SOC. The maximum hysteresis as a function of SOC can be extracted from this test as well.

To compensate for hysteretic effects, a number of approaches have been proposed. The simplest model assumes that the hysteretic effect is of constant, known magnitude, and simply changes sign depending on the most recent value of the battery current. There are therefore only two possible values for the hysteresis voltage, $+V_{h,\max}$ and $-V_{h,\max}$.

For these models, it is important to note that the value for $Vh\max$ is usually dependent upon SOC and temperature. In other cases the value of $Vh\max$ may depend also on the magnitude of the applied current.

$$V_h(t) = sign[I(t)] * V_{h,\max}$$
$$V_h(k) = sign(I(k-1)) * V_{h,\max}$$

In reality, there is a hysteretic loop that occurs as current changes direction. $V_h$ is therefore bounded in the interval $[-V_{h,\max}, +V_{h,\max}]$ but can take on intermediate values. The rate of change of $V_h$ depends upon the applied current.

This model requires implementation of a memory effect when the current is zero, or sufficiently small.

Two commonly used models imply a linear rate of change of $V_h$ as a function of $I$, or an exponential rate of change.

The linear model has the following continuous and discrete-time representations

$$dV_h/dt = k_h I_B$$
$$V_h = clip(-V_{h\max}, V_h, V_{h\max})$$
$$V_h(k) = clip(-V_{h\max}, V_h(k-1) + k_h I, +V_{h\max})$$

The exponential model has a more realistic approach to the hysteresis limit voltage. The hysteresis voltage decays exponentially towards the limit voltage represented by the zero-state hysteresis model.

$$\frac{dV_h}{dt} = \gamma sign(I)(Sign(I)V_{h,\max} - V_h)$$
$$V_h(k) = \exp\left(-\left|\frac{I\gamma\Delta t}{C}\right|\right)V_h(k-1) + \left(1 - \exp\left(-\left|\frac{I\gamma\Delta t}{C}\right|\right)\right)sign(I)V_{h,\max}$$

The constant $\gamma$ adjusts the rate of decay which is proportional to the difference between the current value of $V_h$ and the final value of $V_{h,\max}$ (resulting in an exponential decay function).

The model may be further refined by expanding the maximum hysteresis voltage function to be a function of applied current as well, such that $V_{h,\max} = V_{h,\max}(SOC, T, I)$.

More complex models of hysteresis exist. These have generally been developed to model hysteretic systems where the hysteretic effects are more significant than in battery systems. If warranted, they could be used for lithium-ion battery modeling. One such model is the Preisach model that was developed for modeling of magnetic hysteresis.

In systems with significant hysteresis voltages and/or flat discharge profiles (LiFePO$_4$ is a notable example of both), accounting for hysteresis is required to make useful conclusions about state of charge from terminal voltage.

## 11.4 Coulombic Efficiency

Lithium-ion batteries generally have extremely high coulombic efficiency, greater in many cases than 99%. This means that nearly all the coulombs that are charged into the battery can subsequently be discharged. This means that in many cases it is not necessary to model the coulombic inefficiency to achieve accurate SOC estimation. The relatively small effect will be compensated by other components of the SOC algorithm. However, it may be helpful in some cases to do so. This can be achieved by modifying the behavior of battery model such that the coulombic efficiency is less than 1. This may be done based upon data that gives the battery management system designer a better estimate of the actual coulombic efficiency, or it may be done to provide a bias to the amp-hour integration process to ensure that there is more tendency to underestimate SOC rather than overestimate it. For a number of applications the consequences of underestimation are less severe than overestimation, and this is a simple but effective way to shift the probability distribution towards underestimation.

$$\frac{dSOC}{dt} = \frac{1}{C}\eta I$$

where $\eta$ is the coulombic efficiency and equal to 1 during discharge and is less than 1 during charge.

## 11.5 Nonlinear Elements

Other types of nonlinear circuit elements have been proposed to describe the behavior of battery cells. One is the constant phase element (CPE) (see Figure 11.6). In most cases, the constant phase element behaves similarly to a capacitor. For sinusoidal excitation, the current in a constant phase element always lags the applied voltage by a given phase angle, independent of frequency, that can vary between 0° and 90°. In a true capacitor, this angle is always 90°.

$$Z_{CPE} = \frac{1}{Y_{CPE}} = \frac{1}{Y_0 \omega^n} e^{-\frac{\pi}{2}ni}$$

$$Z_{CPE} = \frac{1}{(j\omega)^n Y_0}$$

For a pure capacitor, $n = 1$, and for a pure resistor, $n = 0$. If for a given CPE, $n$ is close to 1, the CPE can be approximated by a capacitor. In many cases, the "capacitance" in the equivalent circuit of an electrochemical cell is more accurately approximated by a CPE, but a capacitive representation is often used for simplicity.

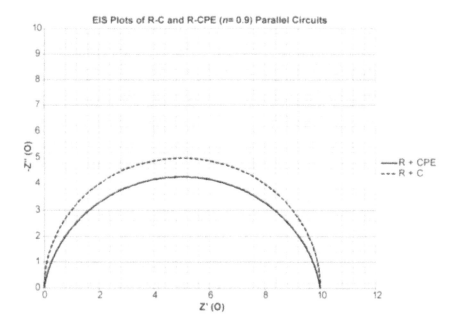

**Figure 11.6** EIS plot of a constant phase element.

In equivalent circuit models derived from EIS, the CPE often appears in parallel with a charge-transfer resistance. The complex impedance plot of this combination is a semicircle with a depressed radius. The angle between the curve and the real axis is $(1 - n) \times 90°$.

Another nonstandard element that is encountered is the *Warburg element*. The Warburg impedance is used to model the electrical effects of diffusion and is a special type of CPE with $n = 0.5$. The impedance of the Warburg element is inversely proportional to the square root of frequency, and has a constant phase shift of 45°.

$$Z_W = \frac{A_W}{\sqrt{\omega}} + \frac{A_W}{j\sqrt{\omega}}$$

$$|Z_W| = \sqrt{2}\frac{A_W}{\sqrt{\omega}}$$

To include the CPE or Warburg impedance in a model for battery management system implementation, a time domain representation (at least approximate) is required. Time-domain representations of the CPE and Warburg impedances are relatively new concepts. Simple linear differential equations can be used to model capacitive and inductive circuit elements which have straight-forward time domain exponential representations and translate easily to discrete-time representations. Closed-form solutions of the CPE and Warburg representations are complicated and a number of approaches rely upon approximations using digital filters [1].

Consider the constant phase element in the Laplace (s) domain. The impedance represents the transfer function of the battery cell, that is, the resulting voltage in response to an applied current. The voltage function in the s-domain is obtained by multiplying the current function by the impedance transfer function. In the time domain, this yields a complex convolution integral. A direct solution of this convolution integral is not generally possible due to the computational complexity, and a simplification is desirable.

Digital filters are an effective representation of such an approximation in the time domain. Digital filter coefficients have the advantage that they can be determined in the Laplace domain through the location of the poles and zeros of the transfer function.

The goal is to obtain a relationship between the voltage and current functions. In the Laplace domain, this is represented by a multiplication of the current function with the filter transfer function.

Figure 11.6 shows the difference in shape of the EIS result between a linear RC circuit using a true capacitance, and a constant phase element for a particular value of the exponent $n$.

Reference [1] gives a useful method for approximating the general transfer function of the constant phase/Warburg element using a digital filter with a variable number of poles and zeros. This approximation gives good correlation in a defined frequency band between $\omega_u$ and $\omega_l$, which is appropriate for most battery system applications. The highest-frequency component can be approximated by $1/t_r$, where $t_r$ is the fastest expected rise time for the battery current; frequencies of 10–100 Hz are high enough in most cases. Although frequency components down to dc will exist in the battery current signal, limiting the low-frequency components to $10^{-2}$–$10^{-3}$ rad/s will provide accurate results for most applications.

Once the upper frequency $\omega_u$ and the lower frequency $\omega_l$ are chosen, the center frequency $\omega_c$ can be found at:

$$\omega_c = \sqrt{\omega_u \omega_l}$$

The approximate transfer function is then expressed as:

$$D_N(s) = \frac{\omega_l}{\omega_c} \prod_{k=-N}^{N} \frac{1+s/\omega'_k}{1+s/\omega_k}$$

where

$$\omega'_k = \omega_l \left(\frac{\omega_l}{\omega_c}\right)^{\frac{(k+N+0.5-n/2)}{(2N+1)}}$$

$$\omega_k = \omega_l \left(\frac{\omega_l}{\omega_c}\right)^{\frac{(k+N+0.5+n/2)}{(2N+1)}}$$

Increasing the chosen value for $N$ ($N$ = 1, 2, 3...) will improve the accuracy of the representation by adding additional poles and zeros (total $2N+1$).

From the pole/zero representation, the conversion to continuous-time domain and the discrete z-domain is straightforward. The implementation of this filter in a digital embedded system consumes limited resources and is well understood.

These approximations can be used to model the CPE and Warburg impedance over a useful frequency range and achieve useful results when linear RC approximations do not provide enough accuracy.

## 11.6 Self-Discharge Modeling

Like all batteries, lithium-ion cells are prone to self-discharge. The very low rates of self-discharge may mean that this does not need to be considered for all applications; however, there are a number of implications of self-discharge on accurate SOC performance. Self-discharge can be characterized using laboratory testing, but the self-discharge rate depends upon temperature and on SOC (cells at high temperature and high SOC will self-discharge faster). Lithium-ion batteries may exhibit self-discharge rates as low as 1% per month, meaning that for battery systems used frequently, there is little benefit from characterizing the self-discharge and implementing a self-discharge model in the battery management system.

Applications involving long periods between voltage measurements and SOC calculations increase the likelihood that self-discharge effects can negatively affect system performance.

Self-discharge occurs through an internal leakage current path and cannot be measured by any external sensor and therefore is "invisible" to the battery management system. Therefore, if it is not acceptable to disregard the self-discharge, it can only be estimated; it cannot be measured.

If an SOC algorithm is established which is robust against the initial estimate of SOC, initial error in SOC due to self-discharge may be large, but will be reduced in time. In many situations, large self-discharge errors are uncommon and the delay in obtaining an accurate SOC is acceptable. However, if this is not the case, a predictive self-discharge compensation can improve system performance.

Self-discharge rates should be characterized through testing. Because self-discharge rates are extremely low, it is important that experiments are designed that are robust against experimental noise. The two most significant factors which affect self-discharge rates most notably are state of charge and temperature. Cells tend to discharge more quickly at high SOC and high temperature.

If the dependence of self-discharge rate on temperature is high, and the battery is used in an environment where the battery temperature can vary considerably, it may be necessary to perform frequent sampling of the temperature to perform a piecewise integration if an accurate self-discharge estimation is required.

In many cases a robust SOC estimator provides the necessary functionality to account for the effects of self-discharge.

## 11.7 Physics-Based Battery Models

### 11.7.1 Doyle-Fuller-Newman Model

The Doyle-Fuller-Newman model is an electrochemical model for lithium-ion cells. The model solves a set of coupled partial differential equations and describes lithium-ion battery behavior. The Doyle-Fuller-Newman (DFN) model is a first-principles model that attempts to model the salient aspects of the actual physics occurring in the battery cell, as opposed to the approximate dynamics of an equivalent circuit.

In this model, the cell is generalized as a two-dimensional problem with one dimension representing the path of the lithium ion through the electrolyte, and the other dimension representing the radial travel of the ion into and out of the active material. The model exists in a two-dimensional domain $(x, r)$. The $x$-dimension spans the distance between the anode current collector, through the negative electrode active material, separator, positive electrode active material and ends at the cathode current collector. The $r$-dimension begins at the center of the particle ($r = 0$) and ends at the nominal particle radius ($r = rp$), which is different in the two electrodes

The model describes ion diffusion, as well as spatial concentration and potential gradients through the anode, cathode, and separator of the cell. The full DFN model consists of a set of six partial differential equations (PDEs). These PDEs model the linear diffusion of the lithium ions through the electrolyte between the two electrodes, radial diffusion in a spherical coordinate system of the ions into the anode and cathode active material particles, and electrochemical reaction kinetics.

A full treatment of the DFN model is beyond the scope of this publication, but has been shown to give good performance in a range of lithium-ion batteries. Despite the fact that it has been shown to accurately model many effects of lithium-ion batteries, the full DFN model is extremely complex and an unlikely candidate to operate in real time on most battery management systems. Although the DFN model is based around a number of physical parameters and quantities (unlike an equivalent circuit model, which uses fictitious circuit elements to only approximately describe battery behavior), these parameters are impractical to measure in operation. A large number of parameters are also required to describe the model that must be solved for in the parameter identification process. This gives rise to the notion of simplified electrochemical models for use in battery management systems.

### 11.7.2 Single Particle Model

The single particle model (SPM) adopts the simplification of modeling both electrodes as single spheres. The state variables of the SPM are similar in nature to those of the DFN model, but the x dimension only spans the electrolyte—the SPM assumes that the variation in states through the thickness of the electrode material is negligible.

The differential equations that form the SPM are partial differential equations (PDEs). Closed-form solutions to PDE are very rare and there are still significant challenges in obtaining a model of this type to operate in real time.

Simplifications can be used to obtain a representation suitable for BMS implementation. The final result must provide a relationship between battery current and terminal voltage.

One simplified SPM [2] gives the following transfer function between voltage and current.

$$\frac{V(s)}{I(s)} = \frac{R_{ct_+}}{a_{s_+}} \frac{1}{A\delta_+} - \frac{R_{ct_-}}{a_{s_-}} \frac{1}{A\delta_-}$$

$$+ \frac{\partial u}{\partial c_{s+}} \frac{1}{A\delta_+} \frac{R_s}{a_s FD_{s+}} \left[ \frac{\tanh(\beta_+)}{\tanh(\beta_+) - \beta_+} \right]$$

$$- \frac{\partial u}{\partial c_{s-}} \frac{1}{A\delta_-} \frac{R_s}{a_s FD_{s-}} \left[ \frac{\tanh(\beta_-)}{\tanh(\beta_-) - \beta_-} \right] - \frac{R_f}{A}$$

The authors propose a multiple pole filter that approximates this transfer function over a frequency range of $10^{-5}$ to $10^2$ Hz, using a series of linear circuit elements.

Like equivalent circuit models, a number of the physical constants (conductivities, diffusivities, and so forth) used for physics-based models are temperature and SOC dependent. Some battery applications operate in a narrow SOC range and the effects of SOC can be neglected.

Other simplifications may be made, including:

- Neglecting the concentration gradient in the electrolyte;
- Neglecting radial diffusion in the particles and assuming that the lithium concentration is equally distributed in the r-dimension;
- Simplifying the reaction kinetics at one electrode (usually the cathode) where they occur much more quickly.

These simplified models (shown in Figure 11.7) are still usually based on coupled PDEs and are nonlinear and an approximation is usually required for implementation in a battery management system.

160    A Systems Approach to Lithium-Ion Battery Management

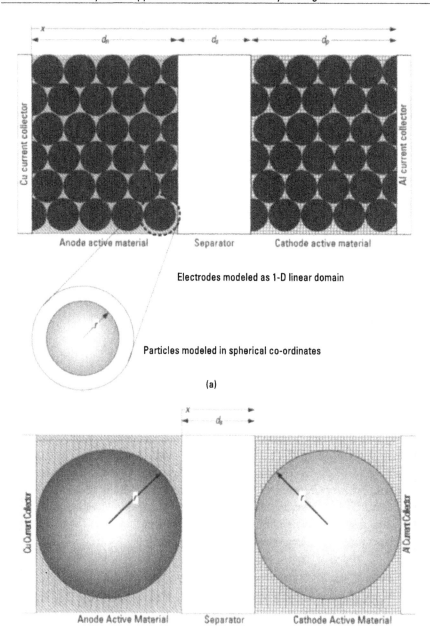

**Figure 11.7**  Comparison of DFN and SPM model domains.

## 11.8 State-Space Representations of Battery Models

State-space representations of battery models are extremely useful in battery management system development. They exist in linear and nonlinear forms, in both continuous and discrete time. They can be used directly with software tools such as MATLAB and Simulink, they can be easily implemented in digital control systems using a discrete time approach, and they can be extended to model nonlinear battery performance quite easily. State-space representations are applicable to equivalent circuit models as well as other types of models as well.

A generic state-space representation of a linear, time-invariant system is

$$\dot{x}(t) = Ax(t) + Bu(t)$$
$$y(t) = Cx(t) + Du(t)$$

The state-space system can be represented equally well in discrete time.

$$x(k+1) = Ax(k) + Bu(k)$$
$$Y(k) = Cx(k) + Du(k)$$

In the above representation, the model has $n$ internal states that are represented by the vector $x(t)$. The values of $u(t)$ represent the inputs to the system, where the values of $Y(t)$ are the system's outputs. $A$ is the state matrix that determines, together with the input matrix $B$, how the state variables change over time. $A$ defines the dependencies of the rates of change of state variables on the values of the state variables themselves, whereas $B$ defines the impact of the system's inputs on the change of state variables. The output matrix $C$ and the feedthrough matrix $D$ define the system's outputs as a function of the state variables and the inputs.

Consider an oversimplified battery model is assumed where there is no internal resistance, polarization or hysteresis, and where the voltage is simply a constant $k$ multiplied by the state of charge.

This system only has a single internal state, the state of charge, which we can denote by $\zeta$. Therefore, $x$ has one element, and $x = [\zeta]$.

In most cases, the only physical quantity contained in the input matrix to a battery system is the battery current. For the simple model discussed here, the rate of change in $\zeta$ is equal to the battery current divided by the battery capacity, $C$.

Therefore, $B = [1/C]$ and $u(t) = [I(t)]$. This gives the linear differential equation $d\zeta/dt = 1/CI(t)$.

In discrete time, $A = [1]$, $B = [\Delta t/C]$, and therefore $\zeta(k+1) = \zeta(k) + I(k)\Delta t/C$.

$Y(k)$ is equal to the single output, the battery voltage, $V(t)$. The output matrix $C$ is equal to $[k]$, and the feedforward matrix $D$ equals the zero matrix, giving $V(t) = k\zeta(t)$, or $V(k) = k\zeta(k)$.

This obviously does a very poor job of representing a true battery as this implies a linear relationship between the state of charge and the battery voltage (in fact, this state space model is appropriate for an ideal capacitor), but it can be improved upon.

The relationship between the state of charge and the open-circuit voltage is nonlinear and therefore a linear system is not capable of modeling this directly. In this case, the output matrix $C$ should be replaced by the nonlinear function $V(\zeta)$.

Adding complexity, the ohmic resistance can be introduced. The ohmic resistance does not introduce additional states, and can simply be represented in the feedthrough matrix $D$.

Because the voltage equals the open-circuit voltage $V(\zeta)$ plus the voltage drop across the ohmic resistance $I(t)Ro$. $D$ is now equal to $[Ro]$ and the state-space model looks like the following.

The addition of the RC elements adds a state variable which could represent either the voltage or the charge on the capacitor.

If the capacitor voltage is chosen as the state variable, for a single RC parallel element, $dV(t)/dt = 1/C[I(t) - V(t)/R]$.

Hysteresis can also be modeled in a state-space representation, requiring additional states to be added. A single-state hysteresis model requires a single state variable to be added, which could simply be equal to the value of the hysteresis voltage.

As is often the case, the values for the coefficients are often dependent upon temperature and SOC. It should be expected that these are nonlinear functions as well.

A state space model incorporating hysteresis, two RC elements, nonlinear SOC-OCV relationship and parameter dependence, and ohmic resistance is given as follows

$$x = \begin{bmatrix} SOC & V_{C1} & V_{C2} & V_h \end{bmatrix}'$$

$$x(k+1) = \begin{bmatrix} SOC(k) + I\eta \dfrac{1}{C} \\ \exp(-\Delta t/R_1 C_1) V_{C1}(k) + R_1 \left(1 - \exp(-\Delta t/R_1 C_1)\right) \\ \exp(-\Delta t/R_2 C_2) V_{C1}(k) + R_2 \left(1 - \exp(-\Delta t/R_2 C_2)\right) \\ \exp\left(-\left|\dfrac{\eta I \gamma \Delta t}{C}\right|\right) V_h(k) + \left(1 - \exp\left(-\left|\dfrac{\eta I \gamma \Delta t}{C}\right|\right)\right) M(SOC, I) \end{bmatrix}$$

$$Y(k) = OCV(SOC) + IR_o + V_{c1} + V_{c2} + V$$

where SOC is the state of charge, $V_{C1}$ is the voltage across $R_C$ element 1, $V_{C2}$ is the voltage across $R_C$ element 2, $V_h$ is the hysteresis voltage, $C$ is the battery capacity, $I$ is the battery current, $\eta$ is the coulombic efficiency, $R_o$ is the ohmic resistance of battery, $\gamma$ is the hysteresis relaxation factor, OCV(SOC) is the open-circuit voltage as function of SOC, M(SOC,I) is the maximum hysteresis as function of current and SOC, and $\Delta t$ is the time step.

## References

[1] Tröltzsch, U., P. Büschel, and O. Kanoun, *Lecture Notes on Impedance Spectroscopy Measurement, Modeling and Applications, Vol. 1*, O. Kanoun, (ed.), Boca Raton, FL: CRC Press, 2011, pp. 9–20.

[2] Prasad, G., and C. Rahn, "Development of a First Principles Equivalent Circuit Model for a Lithium Ion Battery," *ASME Dynamic Systems and Control Conference*, Ft. Lauderdale, FL, 2012.

# 12

# Parameter Identification

Chapter 11 discusses the reasons for having a representative equivalent circuit model and a number of methods to construct these models. The models generally require accurate values for a series of parameters to model a specific battery correctly. This chapter will focus on possible approaches for identifying an appropriate model and for obtaining values for the parameters to accurately represent the battery cells.

## 12.1 Brute-Force Approach

The parameters of the battery model could be approximated in a brute-force method. By observation of voltage/current relationships in the time or frequency domains, one can determine which portions of the response correspond to various elements in the model.

For example, assume the model in Figure 12.1 and the following voltage/current relationship to a step change in current.

The model includes the ideal SOC-OCV relationship, two RC elements, and an ohmic resistance.

If the assumption is made that the time constant of one of the two RC elements is significantly slower than the other, then as time increases, the effect of the ohmic resistance and the faster RC element can be neglected. Conversely, immediately after the application of the current pulse, the effect can be assumed to include only the ohmic resistance. Once the model parameters for the individual elements are determined, the effects of that portion of the circuit can be subtracted from the voltage response curve and the remaining elements (in this case, the faster RC element) determined in sequence. This technique could be

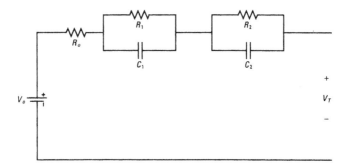

**Figure 12.1** Sample battery model.

used for any number of RC elements. The closer the time constants are to each other, the more likely this method is to introduce modeling errors (see Figure 12.2).

The brute-force approach has a number of drawbacks. The separation of the voltage output into components that apply to model elements may not be obvious to all observers. The application of the separation principle is arbitrary. The method often relies on special current profiles applied to simplify the shape of the response curves. For these reasons, this method is poorly suited to automation and cannot be reliably used for online or adaptive model parameter identification. The parameter values change with state of charge and with battery age; therefore, this method may not be suitable for batteries expected to operate with significant aging effects. For this reason, more automated methods are considered.

## 12.2 Online Parameter Identification

In online or adaptive model parameter identification, the model parameters are determined from automated observation of the voltage and current relationship. No special current profile is required, although it is not unheard of to inject specific current profiles to aid with the parameter identification. Historical battery data is analyzed and the parameters are automatically calculated in approximate real time.

Online parameter identification is valuable as it allows a battery management system to adapt to variations in different batteries that could be introduced by differences in manufacturing as well as aging. In some cases, online parameter identification may be used to create a universal battery management system that can be used with multiple battery types or chemistries.

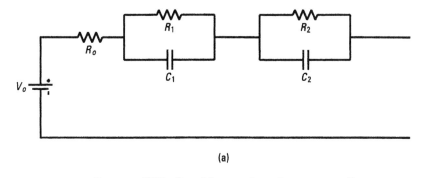

(a)

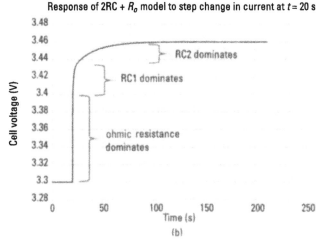

(b)

**Figure 12.2**   Response of two RC and ohmic resistance model to step current pulse.

## 12.3   SOC/OCV Characterization

Many battery models using equivalent circuits require measurement of the relationship between state of charge and terminal voltage, free of any overpotentials. Many overpotentials will relax with time when the battery current is removed, but others are hysteretic in nature and cannot be avoided. Therefore, a method of approximating the overpotential-free voltage is required.

Characterization of the SOC/OCV relationship should be performed at constant temperature to avoid entropic effects.

Use a low charge and discharge rate to adjust the cell state of charge to minimize the excitation of cell dynamics and allow the cells to rest adequately. An initial screening can be performed to understand the maximum relaxation time expected. The required full relaxation time may depend strongly on temperature.

Once an OCV relationship for charge and discharge has been obtained, a neutral curve free of hysteresis can be created between these two assuming that the hysteresis band is of equal magnitude in charge and discharge.

Enforcing a monotonic relationship between SOC and OCV is required to allow equivalent circuit models to work correctly.

## 12.4 Kalman Filtering

A Kalman filter can be used to estimate model parameters. A discussion of using Kalman filters for state of charge estimation will be provided next, but the Kalman filter can also be used for the estimation of battery model parameters.

A Kalman filter can be used as a *state observer* to determine the values of hidden states in a state-space model, which contains both *process noise* and *measurement noise*.

When performing parameter identification, the hidden states to be observed are the model parameters. If the model parameters are accurate, then the predicted voltage response will match the battery response closely.

The nonlinear extensions (EKF and UKF) are nearly always required for parameter update filters due to the highly nonlinear nature of the relationship between the model performance and parameter values.

## 12.5 Recursive Least Squares

A recursive nonlinear least squares method can be used for parameter identification as well. Consider a series of data points from battery testing consisting of voltage measurements during a given current profile as a function of time: $v(I(t), t)$

A battery model can be used to predict the values of $\hat{V}$ given $I(t)$. The magnitude of the error between the model predicted voltage and measured voltage is a measure of the quality of the model. It is desirable to find a set of parameters that minimizes the value of the error between the predicted and measured values of the system states for all points in time.

The accuracy of the model will depend on its structure and its parameters, assumed to be presented by $P = [P_1, P_2, P_3, ..., P_n]$. Assuming that the model structure is appropriate for the battery of interest, the ideal values for $P_i$ are those which minimize the square of error or $(V - \hat{V}(I, t, P))^2$ for all values of $t$.

Define the value $R$ as the sum of all squares of error values for all values of $t$.

$$R = \sum_{t=t_0}^{t=T}\left(V - \hat{V}(I,t,P)\right)^2$$

Initial values for $P$ are chosen using a suitable approximation method. Successive values of $P$ are selected using successive approximation. There are multiple methods for achieving this, but generally relying on computing partial derivatives of $R$ with respect to values of $P$ and updating $P$ achieve a more accurate parameter set. The Jacobian $J$ (derivatives of each term of $R$ with respect to the values of $P_j$) is required to solve for the update vector of $\Delta P$ at each time step.

The simplest method, the Gauss-Newton method, solves for $\Delta P$ using

$$\left(J^T W J\right)\Delta P = \left(J^T W\right)\Delta V$$

where $W$ is a weighting matrix and $\Delta V = (V - \hat{V})$.

Expansions to this method by weighting the observations differently at different time steps and using different methods of solving for the update vector to avoid divergence.

This method is particularly sensitive to large errors in initial parameter estimation. It can be considered effective for determining parameter values during initial cell characterization from data sets, or for online updates of parameters from previous values over the life of the battery where parameter changes occur relatively slowly. A challenge is dealing with the potentially significant changes in parameter values over ranges of SOC and temperature.

## 12.6 Electrochemical Impedance Spectroscopy

Electrochemical impedance spectroscopy (EIS) is a well-known technique in the field of electrochemical cells (both electrolytic and galvanic) used to determine the complex impedance of an electrochemical cell as a function of frequency, over a very wide frequency range, including extremely low frequencies (cycle times in the range of minutes to hours are not uncommon).

EIS measurements are often made at low currents. For large-format systems with high discharge rates, consider increasing the charge and discharge rates to obtain more relevant models and parameters.

Impedance representations from EIS measurements can be used to develop complex equivalent circuits for battery cells.

Plots generated from EIS equipment depict the relationship between the real and imaginary impedance of the battery as a function of frequency. Using experienced observation or the use of special software, an equivalent circuit model can be deduced from the impedance relationships.

Ohmic resistances are generally dominant at the highest frequencies. RC elements create semicircular traces in EIS plots, with constant phase elements in parallel with resistor (R-CPE) are easily identified by depressed semicircles. Warburg elements tend to appear at low frequencies as a straight line.

Figure 12.3 depicts typical EIS results and a possible interpretation.

EIS is of limited suitability for online operation due to the specialized load profile required, but EIS methods can be used with other parameter identification methods for initial parameter determination, combined with online methods for parameter updates.

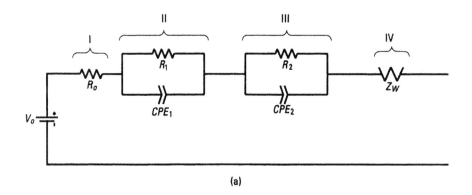

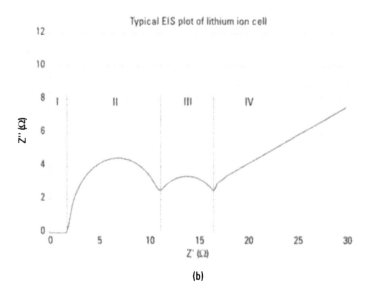

**Figure 12.3** EIS plot for typical lithium-ion battery.

# 13

# Limit Algorithms

## 13.1 Purpose

In most large-format systems, the load device is designed to receive information from the battery management system about the battery's capability for charging and discharging, and to respect the limits imposed upon it by the battery.

As such, the battery management system must implement algorithms to determine appropriate limits in response to the battery condition. These algorithms must be accurate; if they are too conservative, the battery will perform poorly and need to be oversized, and if they are not conservative enough, the battery could be abused by the load under operation. They must respond to dynamics that can evolve at widely varying speeds—a matter of seconds for cell polarization, minutes for changes in state of charge, hours for thermal effects, and years for aging.

In many applications the power limit algorithm is potentially even more important than the state of charge calculation. A hybrid vehicle is a good example of this. The battery in a hybrid vehicle is employed more as a power source/sink rather a significant energy storage device. Because there are multiple sources of energy, the vehicle control scheme will attempt to use electrical energy instead of thermal energy (in the form of fuel) when possible, and to regenerate electrical energy instead of dissipating thermal energy through the friction brakes. As it is assumed that the performance of the vehicle must be independent of the battery's condition, the vehicle control system must know the real-time ability of the battery to provide and accept power very accurately. When the brakes are applied, a power command is being given (the vehicle must be decelerated at a certain rate); it is obviously desirable to store as much of this energy as possible in the battery system, but not at the expense of decelerating too slowly and causing a collision. Therefore, a real-time estimate of the battery's ability to provide and accept power is of utmost importance.

## 13.2 Goals

The limit algorithms will have a number of goals that aim to balance system performance and battery protection.

Maintaining the terminal voltage of all cells in a prescribed range during operation is often selected as the "goal" for a limit algorithm. The word "all" is important—although the total power provided will depend on the aggregate voltage of the entire battery stack, it is the most extreme cell that will limit the performance. This is the challenge associated with computing power limits as opposed to current limits; to compute a power limit, the voltage of both the extreme cell and the entire pack must be predicted.

For most battery cells, a maximum and minimum allowable terminal voltage is specified. The limit algorithm then computes the current or power at which it is predicted the battery terminal voltage will reach these limits.

Other batteries may specify voltage limits that are compensated as functions of applied current and temperature. The longest-term limiter of performance for many battery systems is the thermal capability of the battery system. Thermal limits are usually the slowest to change in response to battery conditions but are equally important to not exceed.

## 13.3 Limit Strategy

A fundamental choice must be made early in the development of both the battery system and load device to determine how the limits should be communicated to the load device.

An important decision at the application level is whether limits should be expressed in terms of current or in terms of power. When calculating the battery response, using a current limit is more intuitive and allows for more control over how the battery will respond, reducing the possibility of overshoot. A power limit is usually more desirable from the load perspective but specifying a power limit requires the limit algorithm to estimate both the current and the voltage at which the limit condition is reached.

Another important characteristic worth capturing is the time-varying nature of the limits. The instantaneous limit may not capture the entire picture of how the load should behave. For a high-performance battery system that is operating at or close to the limits, it is useful that predicted changes to the limits are communicated to the load.

A simple zero-order limit will only provide the instantaneous value of current or power allowed in charge and discharge. A first-order limit will add a rate of change to this. This could be done by providing the rate of change, or perhaps by communicating the value of the present limit as well as the value of the limit at some future point in time. This would provide more information to the load device to allow it to predict future current/power availability and to anticipate changes in limits to prevent exceeding the battery's safe operating area. For systems with high charge and discharge rates at extremes of state of charge, this technique can reduce following errors and allow the load to better utilize the battery, especially if there is a significant time delay between limit commands being issued and the load responding, or the slew rate of the load current is limited. Extensions to higher-order limits can be made with diminishing returns.

A common strategy to prevent oscillations between the battery limits and load current is to impose a restriction on the allowable slew rate of the battery limits, except in the case of the most severe events. If the rate of change of the battery limits is slower than the response time of the load and all associated communications and delay loops, then issues with overshooting the new limits will be eliminated. This requires careful analysis to ensure that safety issues do not arise, but in general this strategy can eliminate concerns with load transient response to changes in available battery power.

## 13.4 Determining Safe Operating Area

The goal of the limit function of a battery management system is to ensure that the battery is not operated outside of its safe operating area. Like many battery characteristics, defining the safe operating area is not necessarily a black-and-white determination as the degree of safety will vary continuously as battery parameters change, but, in general, a continuous set of battery conditions including cell voltage, current, temperature, and SOC should be established that is considered acceptable from the safety, lifetime and performance standpoints.

The application must be well defined. A set of limits established for a battery cell may be perfectly safe and provide good performance in one application but lead to unacceptably fast degradation with cycling in another application. Comprehensive experiments, examining the critical variables of operating and storage temperature, cycle and calendar life, charge and discharge rates, and minimum/maximum SOC operation window should be performed to create a general-purpose life model. This can be augmented with physics-based models of degradation phenomena and cell disassembly and inspection to verify models and to link internal physical effects with externally observable electrical and thermal macroscopic behavior. A tool such as this will provide the battery system and battery management system developers with a predictive tool that can be used for each new application.

To establish an effective limit algorithm, the following information must be available:

- *Temperature:* What are the maximum limits of temperature for normal operation, which may be different for charging and discharging? What are the maximum permissible rates for charge and discharge as a function of temperature? What are the critical temperatures where thermal runaway and other damaging effects begin to occur? What operating temperature window will provide the desired cycle and calendar life?
- *Voltage:* What are the maximum and minimum recommended cell voltages to prevent cell degradation, and what are the maximum and minimum voltage limits to avoid cell damage or dangerous conditions? This information should be available from the cell manufacturer, but can be independently verified through testing.
- *Resistance:* What is the expected dc resistance of the cell at the expected range of temperature, SOC, and age to be encountered?
- *Rate:* What is the maximum recommended charge and discharge rate as a function of temperature, SOC, and internal resistance?
- *SOC:* What is the recommended SOC interval for the application in which the lifetime requirements can be met?

Manufacturers of lithium-ion cells may or may not have comprehensive answers to all of these questions, especially for application-specific use cases.

## 13.5 Temperature

The capabilities of any battery, lithium-ion or otherwise, will be strongly dependent upon temperature.

Many manufacturers prescribe maximum operating temperatures beyond which the battery should not be operated. As battery temperatures approach these ranges, the battery management system should reduce charge and discharge limits to zero (see Figure 13.1). These temperature limits are generally different for charging and discharging.

If the cell manufacturer's recommendations are to be strictly followed, the battery management system should act on the extreme values of measured battery temperature. For example, charging is often inhibited at low battery temperature. It is critical that the minimum measured temperature be used

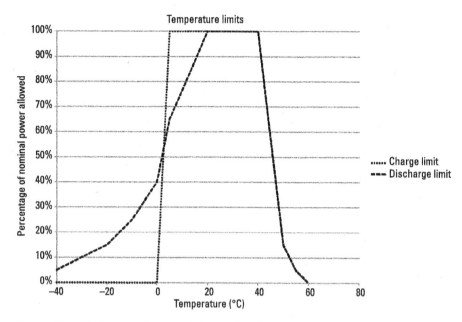

**Figure 13.1** Discharge and charge limits as a function of temperature.

rather than an average temperature to ensure that no cells are operated outside their recommended ratings. Additionally, if it is possible that the true extremes of temperature may lie outside the range of measured values (due, for example, to the temperature sensors not being able to measure the hottest and coolest parts of the cell), or if the measurement errors at the temperatures of interest are significant, additional margins of safety must be included.

Because the limited operation of the battery system is undesirable, it is expected that the battery management system will compute the least restrictive limits possible which maintain the safety of the battery without unnecessary power reductions.

Steps can be taken to minimize potentially exceeding the upper temperature limit due to battery self-heating by rolling back the limits gradually as battery temperatures increase. If the roll-off is too abrupt, or occurs at too high a temperature, it is possible that the maximum recommended temperature may be exceeded if the battery experiences high current operation while already at an elevated temperature. Under no circumstances should operation be permitted at very high temperatures where the margin of safety to the temperature at which thermal runaway can begin is inadequate, or where other types of damage can occur.

Temperature affects the battery's internal impedance, polarization and hysteresis characteristics significantly. The impedance of most lithium-ion batteries increases significantly at temperatures below 0°C, and may increase an order of magnitude between room temperature and low temperatures (−20°C to −30°C). In this range, the power available to the load without experiencing extreme voltage excursions will be reduced severely. While a closed-loop limit based on cell voltage would be capable of reacting to the increased impedance, generally a simple feedforward limit using a temperature-based lookup can improve limit performance.

A thermal model can be used to account for thermal time constants between internal heating of cells and temperature sensors. A simple lumped parameter thermal model can usually accurately predict temperature rises and compensate for delays and errors in temperature measurement. Figure 13.2 shows such a model.

The thermal model can usually consist of a lumped thermal RC network, where the heat capacity of each component and thermal resistances between them are modeled as single elements. Heat sources will consist of the batteries themselves as well as any externally applied heating, and heat removal will take place to both the ambient environment as well as to any active cooling systems which may exist.

Endogenous battery heat sources consist of:

- Joule or ohmic heating that occurs as a function of current flow. An equivalent thermal resistance (ETR) can be determined from a temperature rise test and used to calculate joule heating using $I^2R$. This resistance may not necessarily be the same resistance used for electrochemical/equivalent circuit modeling. The ETR will be a function of temperature, SOC, and age and must be tracked dynamically if the thermal model is to remain accurate.

- If a significant voltage hysteresis exists between charge and discharge, then energy is lost in the form of heat proportional to the size of the

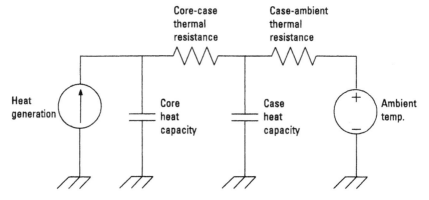

**Figure 13.2** Sample thermal circuit of battery system.

hysteresis loop traveled. If a voltage offset $\Delta V$ exists during cycling at constant current I, then energy is lost at a rate equal to $I\Delta V$ during the hysteresis loop. The proportion of heat lost during charge and discharge is not necessarily equal, and should be evaluated by testing. Additionally, the magnitude of $\Delta V$ will possibly depend on the battery current as well as the usual parameter space of SOC, temperature, and age.

- Entropic heating/cooling: The charge and discharge reactions often have a non-negligible reversible thermal component; charging may be exothermic (releasing heat, which warms the batteries) while discharging may be endothermic (consuming heat, which cools the batteries). This can be tested by one of two methods: either by subtracting the ohmic heating effects from the thermal rise during testing, or by calculating the change in entropy analytically using thermodynamic relationships.

A number of temperature measurements are available for limit calculations. Normally the two extreme temperatures (which should be approximately equal in a well-designed battery system that is functioning nominally) should provide a worst-case condition for temperature effects. For effects with high sensitivity such as inhibiting low temperature charging, operating close to temperature extremes, extra margins of safety, as well as the possible effects of measurement uncertainty should be included in the limit calculation to ensure that the cells are not abused.

## 13.6 SOC/DOD

The capability of a battery to provide and accept power varies with SOC/DOD at a given temperature.

A useful test to determine the battery's power capability at various operating points can be performed using a constant voltage/constant current cell cycler capable of supplying enough current to drive the cell to the maximum recommended charge and discharge voltages across the full range of cell state of charge (for low impedance cells designed for high power applications, this could be a very high current).

The cell should be set to the desired test conditions (SOC and temperature) and allowed to stabilize both thermally and electrochemically.

The cycler should then be set to operate in "constant voltage" mode to push the cell to the limit voltage. A current limit should be in place to prevent the cell from exceeding the charge/discharge current ratings and temperature monitoring should also be in place. The current will initially rise to a value limited by the ohmic resistance alone, and will fall gradually as the cell dynamic overpotentials are excited.

The power at various times since starting the voltage command pulse can be calculated as voltage × current, and therefore an estimate of power capability for pulses of various lengths can be established.

The test should be repeated in charge and discharge and the cell should be returned to the target SOC and temperature before each pulse. Due to hysteresis and polarization, the measured voltage on cells brought up to the target SOC from 0% SOC may differ significantly from cells discharged to the target from 100%. Accurate cycling equipment is important to achieve good results.

Repeating this test across the range of SOC and temperature expected during operation will provide a very useful estimation of battery capability in charge and discharge. This can be useful in establishing a predictive limit baseline for the recommended maximum current as a function of SOC; however, it will suffer from a number of drawbacks, including:

- If the SOC is inaccurate, there will be consequent errors in the predictive limits.
- As the cell's overpotential dynamics change as the battery ages, specifically as the resistance increases, the limits will not be sufficiently conservative to protect the battery even if the SOC is accurately known.

Therefore, limits as a function of SOC should be used as a feedforward or predictive limit in conjunction with other information to maximize performance.

With SOC estimation schemes where the error bounds or confidence interval of SOC is known, the error bounds may be taken into consideration as well.

Beyond the power availability, for many applications the SOC window of operations is reduced significantly from 0% to 100% to provide the necessary cycle life. Hybrid vehicles and other peak power applications are good examples of where this is needed. In these uses, even if it is safe to do so, battery operation must be limited outside this range of SOC. Therefore, an application-specific limit (as opposed to a limit related to the battery capability) can be imposed to prevent excursions outside of this window. A simple linear taper of charge at high SOC and discharge at low SOC is adequate for most applications. Normally this limit strategy is combined with algorithms in the load device which determine predictively how the battery should be used as a function of its reported SOC.

## 13.7 Cell Voltage

Measured cell voltages provide the best opportunity to generate responsive limit information. An appropriate response to cell voltages approaching their limits is the most time-critical piece of battery management system functionality that will be exercised in normal system operation. As many cell chemistries approach their limits of full charge and discharge, the relationship between cell voltage and state of charge becomes very steep and voltage limits may be rapidly exceeded if high rates of charge or discharge are sustained. A component of cell terminal voltages is also due to overpotentials caused by cell excitation, so cells that are exposed to excessive charge or discharge rates (especially at low temperatures or end of life where impedances are highest) can cause excessive voltages to occur. Even if the cell state of charge remains in an acceptable range, overpotentials can cause cell degradation to occur; this limits the safe operating area in both voltage and SOC, and leads to separate failure modes of both overcharge and overvoltage.

Safety can be achieved by disabling charge at high cell voltages and discharge at low cell voltages, but these types of rapid adjustments in battery current are not acceptable in most situations. An accurate battery model will give a good estimate of the maximum overpotential that will be produced as a function of current, combined with accurate SOC (or at least OCV) estimate will give a maximum steady-state current in both charge and discharge. Unlike for state of charge estimation, which needs to operate on the cells with the largest allowable charge and discharge capacities, the cell voltage limits model needs to operate on the cells with the highest and lowest terminal voltage. Accounting for measurement errors in both current and voltage will improve safety margins as well.

A simple dynamic control loop can be used near the edge of the safe operating area. If the relationship among current, SOC, and total overpotential is known, along with the relationship between SOC and OCV, this information provides not only the maximum allowable current at the present operating point, but also the rate of change of this limit with respect to time. For systems operating at high power-to-energy ratios, this can allow for a very responsive control when operating near the limit.

As usual, the above quantities and relationships are dependent upon temperature as well, but the rate of change of battery temperature can generally be ignored in most short-term cases where operating near the battery limits.

A common problem presents itself as the battery approaches the end of discharge and the discharge limit reduces, and the load, in turn, reduces the discharge current, causing cell voltage depression to relax leading, in the case of an inaccurate or simplified model, to recovery of the limits, allowing higher discharge currents. An oscillating condition can result, which can lead to an unstable performance.

Adding some hysteresis, either discrete or continuous to cell-voltage based limits, can prevent oscillation between the battery and the load due to this effect.

## 13.8 Faults

In the event of noncritical battery failures or battery management system faults, it is often desirable to reduce the battery charge and/or discharge rates to improve safety but to provide some limited performance. Reduced rates, more limited voltage and SOC windows, and more sensitivity to temperature can improve the safety of the system in the event of compromised sensors or measurement, unbalanced cells, or potential errors detected in SOC estimation.

Reducing limits is warranted if a significant cell imbalance exists. A forced overdischarge, leading to cell reversal, can be created if a large-format battery system is discharged when the lowest capacity cell reaches 0% SOC. The voltage can fall very quickly and a cell can enter cell reversal with little warning. Either a cell voltage limit will activate, causing abrupt changes in limits, or the detection will be too slow and overdischarge will occur. In applications in which the battery is fully utilized, for this reason it may make sense to taper allowable limits as the battery approaches 0% and 100% SOC.

## 13.9 First-Order Predictive Power Limit

Assume that the battery cell is modeled by an ideal battery in series with a single internal ohmic resistance. A minimum and maximum acceptable terminal voltage, $V_{LIM,MAX}$ and $V_{LIM,MIN}$, are established for the cell. The limit condition is reached when $V_{OC} + IR_o = V_{LIM,MAX}$. Solving for $I$,

$$I_{MAX,CHG} = \frac{V_{LIM,MAX} - V_{OC}}{R_o}$$

$$I_{MAX,DIS} = \frac{V_{LIM,MIN} - V_{OC}}{R_o}$$

The power at this point is equal to $V_{lim} * I_{MAX}$, or $(V_{oc} + IR_o)I = IV_{oc} + I^2 R_o$. $V_{oc}$ is assumed to be a function of SOC. $Ro$ may be looked up as a function of SOC and temperature, it may be calculated dynamically, or it may be a combination of both.

Ignoring polarization and other dynamics, this provides a power and current limit for a battery that will maintain the terminal voltage between $V_{LIM,MIN}$ and $V_{LIM,MAX}$.

The cell's open-circuit voltage and the resistance $Ro$ can be functions of SOC and temperature. Therefore, if SOC and battery temperature are known, this first-order limit can simply be looked up.

If the resistance value is conservatively chosen to account for the maximum overpotential due to all internal cell dynamics, this power-limit algorithm will provide good performance for preventing safety violations. The power limit may be unnecessarily conservative in an application that switches rapidly between charge and discharge or in cells with very slow internal dynamics.

## 13.10 Polarization-Dependent Limit

If a battery model such as that described in Chapter 11 is operating in real time, the total polarization due to the RC elements will be known. This information can be used to further refine the limit algorithm.

This can be particularly advantageous in an application in which the battery must rapidly reverse between charge and discharge, the currents are high, and/or the total energy capacity is low.

For higher-order time-based limits, the battery model can be operated in a predictive manner to determine the length of time for which a given current can be discharged or charge before a limit voltage is reached.

## 13.11 Limit Violation Detection

The battery management system is required to detect violations of the published limits and to respond to this potentially dangerous condition. However, in the majority of cases, the battery management system has no authority to "enforce" the limits by restricting battery current or power other than to open contactors and disconnect the battery from the load. Therefore, a nuisance trip of this fault condition must be avoided. The limits can be exceeded due to either a load device that is malfunctioning, in which case the problem could be persistent, or due to a slow response to a transient condition that is likely to correct itself as the load responds. A simple detection strategy that opens contactors as soon as limits are violated in the slightest way will likely create many more problems than it prevents.

A "leaky bucket" integrator strategy is a useful technique for detecting limit violations. The battery management system integrates the current or power that is in excess of the permitted limit. This integrated error decays over time, either at a fixed rate, or at a rate proportional to its magnitude (exponential decay). When the integrated error reaches a predefined value, the battery will be disconnected. This prevents small errors (which are unlikely to cause a safety issue) from accumulating and causing a trip, and makes the system respond quickly to large errors.

Emphasis on responding to higher magnitudes of limit violation can be achieved by integrating $(I - I_{LIM})^2$ instead of $(I - I_{LIM})$; this more closely mimics the behavior of a thermal fuse when the limit is exceeded.

## 13.12  Limits with Multiple Parallel Strings

In systems employing multiple parallel strings, which may be online or offline depending on the status of the individual battery subsystems and which may have differing states of charge, and perhaps even different temperatures and ages, the problem of computing limits for the parallel combination of these battery systems grows in complexity.

In parallel string systems, each parallel string is usually capable of only providing a fraction of the system's rated power. Not all strings may be simultaneously online.

If the battery strings are simply connected in parallel, it is difficult to simply add all of the limits together as the strings may not share load current equally depending on their respective impedances and states of charge. Once again, an accurate cell model is important to understand what each battery string's contribution will be and to determine an overall limit.

# 14

# Charge Balancing

To maintain battery performance over a long service life in a large-format battery system, it is usually necessary to implement a charge balancing strategy to account for differences in cell performance.

An effective cell balancing system maintains the desired level of battery performance throughout the life of the battery with an appropriate safety margin, without adding excessive cost, weight, or complexity. To design appropriate cell balancing, the batteries themselves must be fully understood. Due to their high coulombic efficiency, lithium-ion batteries do not "self-balance" like many other types of batteries. Without proper management, battery imbalance will not correct itself over time. A number of system and battery cell parameters are critical to understanding the cell balancing function.

Understanding the expected differences in cell capacity is important. In the absence of charge-transfer balancing, the capacity of the entire series-connected string will be limited to the capacity of the lowest capacity cell. If a charge-transfer capability exists, it can be used to transfer energy from high-capacity cells to low-capacity cells during the discharge to increase the effective capacity of the battery.

Knowing the real-time difference in state of charge between cells is also required for effective balancing.

Self-discharge rates and differences in the self-discharge rate between cells are one of the primary drivers for sizing the balancing circuit. Battery self-discharge is, in general, undesirable and should be minimized wherever possible by the cell supplier. Differences in self-discharge rates will stem from either variation in the manufacturing process or are due to the presence of defects in the cells. A diagram explaining the types of variation that may occur is shown in Figure 14.1.

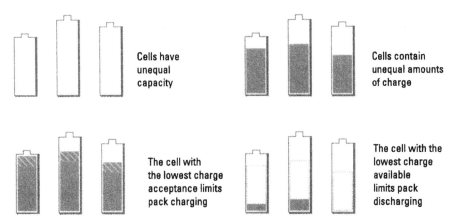

Figure 14.1  Cell variation and need for balancing.

The cell balancing system will be designed to meet one or more of the following design goals:

- *Minimize differences in charge between cells:* The effective capacity of the pack is reduced by the difference in charge between the most and least charged cells.
- *Maximize available battery power:* Cells at different SOCs have different power capabilities due to the influence of SOC on cell impedance. If cells drift to high or low SOC, they will limit the battery power capability as well as the energy capacity.
- *Maximize available battery energy:* If cells are not of equal capacity, cells with higher capacity will still contain useful energy when the lowest-capacity cells reach full discharge, but because this energy cannot be extracted without overdischarging the "smaller" cells, the energy will be stranded. If charge can be moved from the larger cells to the smaller ones, this stranded energy can be recovered.

## 14.1 Balancing Strategies

The simplest cell balancing strategy simply aims to prevent cells from diverging in state of charge over time. Cells with higher self-discharge rates must receive more charge current (or less discharge current) to compensate for the self-discharge rate to prevent divergence. While this technique does not extract maximum performance from the battery system, due to stranded energy as a result of capacity imbalance, it will ensure that battery capacity is not significantly reduced below that of the lowest-capacity cell due to severe battery pack imbalance. Selective charging or discharging is used to adjust the state of charge of cells so that the SOC of all cells does not diverge which will lead to a further reduction in available capacity.

If it is assumed that not all cells have equal SOC and capacity, then two extremes are possible. A battery management system can perform no equalization, in which case at any operating point, the discharge and charge capacities of the entire pack will be equal to the lowest available capacity of the individual cells. Conversely, the battery management system could perform active equalization at a high rate, in which case the entire energy content of each cell can be utilized. A number of intermediate possibilities with different costs exist between these two extremes.

There is a theoretical maximum useful balancing capability, beyond which no additional benefit is obtained from higher balancing currents. This occurs at the point when all useful energy imbalance has been extracted during the time expected for a full discharge.

The creation of SOC imbalance and capacity divergence is a time-dependent process and proceeds at a slow rate. If extremely high rates of self-discharge exist, this is usually a warning sign of a more serious internal short circuit exists and battery operation should be inhibited. Therefore, it is easy to overestimate the need for balancing capability.

## 14.2  Balancing Optimization

Because the balancing circuits must be replicated for each cell, increasing the battery management system balancing capability has a high potential to add cost to the system due to the large multiplier of components which are needed. Therefore, it is important to size the balancing circuitry appropriately to ensure that the battery performance is maintained by preventing divergence without unnecessary cost.

Like any optimization, a relevant cost function must be defined to obtain useful results. If the penalty for inadequate balancing is reduced available capacity from the battery system, the cost of this imbalance is the cost of adding additional battery capacity to compensate for the lost available energy. For example, a system with a 2 Ah capacity penalty due to battery imbalance must provide an additional 2 Ah of capacity in order to compensate.

The selection of the balance point is an important characteristic. The balance point is where, if the pack is perfectly balanced, all the cells (which are of different capacities) are at the same SOC. For example, if the balance point is 100%, when cells are perfectly balanced, they all reach 100% SOC at the same time during charging, but they will diverge in SOC during discharge (smaller capacity cells dropping faster than others). If the purpose of the battery is to provide approximately equal capability for charge as well as discharge (as in the case of a hybrid vehicle or some energy storage peak-shaving applications), the balance point may be better located near 50% SOC.

Where to place the software logic for the balancing decision-making in distributed systems is another important consideration. As information about all individual cell SOCs is needed for balancing decision-making, placing the balancing algorithm in the master device is likely the most appropriate approach as the condition of all cells is needed to determine the appropriate balancing activity for each cell.

Determining the way in which the cell balancing circuitry should operate is critical to maintaining the state of balance. If the balancing decisions are made continuously throughout the operation of the battery system using individual cell SOCs, then the assignment of how to balance the cells can be made nearly continuously. In many cases, however, the computational resources for running a full SOC estimation on a large number of individual cells are usually not cost-effective. Therefore, if the application permits, it may be more useful to wait until a more straightforward measure of relative state of charge is available, for example, after a long period in which the battery is open-circuit and battery currents are zero, when open-circuit voltages are available for each cell.

If the nonhysteretic overpotentials are assumed to be relaxed and the hysteresis is roughly equal for all cells, absolute or relative SOC information can be obtained on a cell-by-cell basis during this relaxed state with much less computational effort. Accuracy can be further improved in the presence of measurement noise by obtaining measurements over a longer period of time at open-circuit, since rapid responses to transient conditions are unneeded. Assuming that the differences in SOC are small, a simple first-order linear approximation can be used to approximate the relative SOCs and create the balancing assignments.

The relative SOCs should be transformed to a number of amp hours to be discharged from each cell. In the case of dissipative balancing using a fixed resistor, the balancing current is given by $V_{CELL}/R_{BAL}$. As the cell is discharged, $V_{CELL}$ will drop, decreasing the balancing current. In many cases the error created by an assumption of constant balancing current is negligible, and a time duration to balance can be assigned for each cell. In the case of low-power balancing circuits, these times may be very long (hours to days) and will likely need to be maintained in nonvolatile memory across cycles, while decrementing the counters as cells are being discharged. Attaching a maximum time to balance to prevent "stale" data from being used for a long period of time during which the state of balance has changed is an effective way to prevent balancing errors in systems that do not have the opportunity to balance often.

In many cases, the battery management system can only perform cell balancing in some sort of active state; control power may be required, leading to other system requirements that may even include closed contactors to provide control power from the high-voltage battery stack itself. In this situation balancing may cause loss of battery energy and therefore a trade-off exists between performing fully accurate cell balancing or minimizing wasted energy. If the consequences of imbalance are potentially severe, consider modifying the operating states and power architecture of the battery management system and battery pack such that the battery system can be balanced with no impact to other system components and with minimal energy consumption.

## 14.3 Charge Transfer Balancing

Charge transfer balancing provides a method for transferring charge from one cell to another. Not all of the energy is therefore lost. The principal advantage of this technique is that when the available energy from the most discharged cell reaches zero, with a dissipative balancing system, all the energy in all other cells is stranded, whereas with charge transfer balancing, this energy can be partially utilized. This can increase the overall energy capacity of the battery pack by an amount up to the energy imbalance between cells due to both capacity and SOC inequality.

### Example 14.1

Consider a 100s configuration of $15.0 \pm 1.0$ Ah cells with battery capacity equally distributed between 14.0 and 16.0 Ah.

With dissipative balancing, the maximum battery capacity is 14.0 Ah (capacity of the lowest cell), if all cells reach 0 Ah available at the same point in the discharge cycle. Errors in relative SOC will further reduce capacity.

With charge transfer balancing, the maximum battery capacity is theoretically 15.0 Ah (the *average* capacity of the cells), assuming energy can be moved arbitrarily from any cell to any other cell with 100% efficiency at a rate fast enough to ensure the energy is available during the discharge cycle.

Although the increased efficiency and ability to improve the available capacity are features that make charge transfer balancing desirable, careful consideration must be applied. In many situations, charge transfer balancing does not provide enough benefits to justify its inclusion. The following points should be analyzed in making this decision.

- *Difference in cell capacities and self-discharge rates:* All modern battery manufacturers are pushing to reduce the variation between cells. As such, for most high-quality cells, the differences between individual cells are very small. Because the best possible improvement in available capacity from the battery is the sum of the residual energy in all cells once the most-discharged cell reaches 0% SOC, if the cells are close in capacity, the amount of energy that can be transferred will be very minimal.

- *Expected length of charge and discharge cycles and overall battery management system balancing duty cycle:* If the battery is discharged very quickly, the energy will need to be correspondingly transferred more quickly to make use of the energy during a cycle, meaning that the balancing circuit will need to be sized much larger. For example, a 100-Ah battery pack with an expected 2% capacity imbalance with an expected 1-hour discharge rate will need to move 2 Ah/h or 2A of charge transfer current to use all of the energy in the high-capacity cells. If the expected discharge time is cut in half to 30 minutes, the balancing current will need to double to achieve the same result.

Charge transfer balancing can be achieved in different ways. A number of fundamental topologies are possible to implement charge transfer between individual cells.

In a master-slave battery management system topology, it is important to decide whether charge can be transferred only within a single slave device, or whether the battery management system needs to be capable of transferring change between all cells in the system.

Because the cells are all at significantly different potentials, methods of charge transfer balancing may potentially involve components rated for high-voltage switching and/or isolation ratings. Because this can increase the cost of the battery management system considerably, it is desirable to find ways to reduce the potential differences for the balancing circuits.

In systems with an isolated low-voltage system connected to earth ground, it is possible to transfer balancing charge across the isolation barrier to an earth-ground referenced system (the 12-V electrical system in an electric vehicle is a good example). Through the use of a flying capacitor or a transformer balancing circuit, energy can be transferred across an isolation barrier while maintaining the integrity of the barrier.

### 14.3.1 Flying Capacitor

The flying capacitor consists of a capacitor that can have one or both of its terminals connected to multiple devices for the purpose of charge transfer or measurement. A flying capacitor circuit can be used to make measurements or transfer charge across an isolation barrier if it can be ensured that the capacitor is not connected to both sides of the barrier at the same time. A representative schematic of a flying capacitor circuit is shown in Figure 14.2.

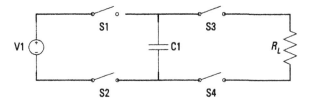

**Figure 14.2** Flying capacitor circuit.

When closed, switches S1 and S2 connect the capacitor to voltage source V1. Capacitor C1 will then charge up to V1 through resistor R1. Switches S1 and S2 can then be opened and S3 and S4 closed to connect the charged capacitor to load resistor $R_L$. The energy in the capacitor is transferred to the load resistor.

Additional pairs of switches could be added to connect the capacitor to any number of sources or loads. The voltage source and load resistor could be replaced by two battery cells, and by closing S1 and S2 in alternation with S3 and S4, energy from the cell with higher potential is transferred to the cell with lower potential.

The load resistor could be replaced by a voltage measurement circuit. This allows the measurement of a multiple voltages across an isolation barrier with a high level of safety. It allows multiple measurements to share a single high-accuracy measurement circuit and analog-to-digital converter. Different voltage dividers could be used between the switches and the voltage sources allowing the measurement of many different voltage ranges with the same circuit.

Limitations of this circuit for charge transfer requires that the load voltage be lower than the source voltage; it cannot be used to move charge from a lower potential to a higher potential. It is not immediately obvious that this may be desirable, but if cells are different in capacity, then a cell at a lower voltage may in fact have higher available energy if it has a higher capacity. Also, the amount of charge which can be transferred is proportional to the voltage difference which may make fine-tuning of the balance difficult as the ability to move current is impeded as the voltage of the source and sink become nearly equal.

A safety issue can exist if switches are closed in the incorrect sequence leading to two switches being closed at the same time, or in the case of solid state switches, inadequate time allowed for the switches to fully open before closing the next one.

Practically speaking, the switches can be implemented using electromechanical relays, optoisolated relays, or transistors. If isolation is required, then an isolated switch or isolated control line is required to ensure the reference potential of the control circuit is isolated from the voltage that is being connected across the capacitor.

A low-leakage capacitor should be selected for this application, as any self-discharge of the capacitor will reduce the effectiveness of the circuit.

A flying capacitor balancing system offers a very low cost option for charge transfer balancing (capacitors are much less expensive than inductors or dc-dc converters) with the disadvantages that the system is much less effective at balancing when differences in voltage are small.

Allowing for connection of the capacitor to any arbitrary cell requires that a switch capable of blocking the full pack voltage is used on both terminals of the capacitor and that a large matrix of switches, capable of connecting the capacitor to any cell in the pack, is created. The cost of this type of arrangement is extremely high and rises as the required isolation voltage increases.

Another arrangement is to allow charge shuttling only between adjacent cells. In this way each switch only needs to be capable of blocking the voltage of a single cell. A simple transistor will serve this purpose quite adequately, but now an individual capacitor is required between each cell.

Some cell monitoring ICs, such as the Atmel ATA6870, support a flying-capacitor balancing circuit between each adjacent cell.

### 14.3.2 Inductive Charge Transfer Balancing

Storing energy in a capacitor requires a voltage differential to charge or discharge the capacitor. This presents a challenge when many cells have nearly the same voltage and the goal is to still move charge between cells at a meaningful rate. Using an inductor as the energy storage element can improve the system performance significantly.

A topology for inductive charge balancing is shown in Figure 14.3. Charge can be moved from cell $n + 1$ to cell $n$ through the following process. Switch S1 is closed allowing current to flow from cell $n + 1$ through inductor L1, thus storing energy in the inductor. When the switch opens, the current flows through cell $n$ and diode D1, transferring the energy to cell $n$. This method does not require that cell $n + 1$ be at a higher voltage than cell $n$, nor is the ability to transfer energy significantly limited by the voltage differential. The current waveform in the inductor is a sawtooth shape. The inductor size, switching frequency, and transistor should be chosen to achieve the desired balancing current without excessive inductor saturation. The losses in this circuit are limited to the parasitic resistance of the inductor, the switching and conduction losses in the transistor and the reverse recovery and the conduction losses in the diode. Due to the relatively low voltage of a single lithium-ion cell, a Schottky diode with low forward voltage drop is recommended for this application. This is, in most cases, a high-efficiency energy transfer method. The balancing current is limited only by the selection of the components involved.

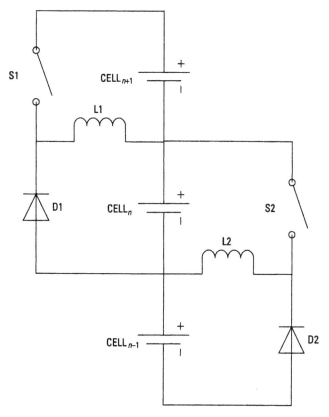

**Figure 14.3** Unidirectional charge transfer balancing circuit.

This topology is only capable of moving energy from a more positive cell to a more negative cell, but the addition of a second inductor and transistor can create a circuit capable of bidirectional energy transfer, as shown in Figure 14.4. Energy transfer from cell $n$ to cell $n + 1$ is accomplished by first closing switch S2 and allowing current to flow in inductor L2. When the switch is opened, current flows through cell $n + 1$ and diode D2, transferring energy to the more positive cell. This requires double the number of switches and inductors. The circuit complexity can be minimized by using the internal body diode of a MOS transistor for each switch.

Because this circuit is only capable of moving energy from one cell to its adjacent cells, if energy is transferred across a large number of cells, the efficiency of this transfer is significantly lower. In the case of large-format systems with hundreds of cells, even a 99% efficient transfer could leave very little energy if the transfer distance is long. This method would be most effective if the weaker and stronger cells are evenly distributed throughout the battery system and the transfer distances can be minimized.

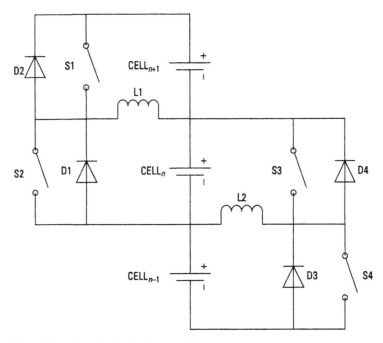

**Figure 14.4** Charge transfer balancing circuit.

An effective control strategy for this type of balancing scheme must consider two important aspects; both the relative energy content of each cell, as well as the efficiency of the transfers between various pairs of cells. Moving energy from the most charged cells to the most discharged cells may not be the ideal choice if significant energy is lost in moving the energy across a long path; it may be more beneficial to select a balance strategy that maximizes the available energy considering the energy loss.

**Example 14.2**

A battery pack with $n$ cells each with available energy $E(n)$. An efficiency function for transferring energy from cell m to cell n is defined as $\eta(m, n)$. The efficiency calculation in many topologies is not commutative [i.e., $\eta(m, n)$ is not equal to $\eta(n, m)$]. The maximum available energy is $\Sigma\ E(n)$ for all $n$ if it is possible to extract all the energy from every cell. However, the available energy without charge transfer is min $(E(n) \times n)$ (assuming that the cell with the lowest available energy limits the total discharge). Energy availability is highest when the difference between min $E(n)$ and max $E(n)$ is minimized. Charge transfer balancing can improve energy availability by performing energy transfers to equalize cell energy. In the circuit shown in Figure 14.4, the inefficiency scales in a roughly linear fashion with the distance that the cells are seperated from each other in the series chain.

### 14.3.3 Transformer Charge Balancing

Expanding on the idea of using an inductor as the energy storage element, a second winding can be added to the core, creating a transformer that can be used to transfer energy between individual cells and a module, the entire battery stack or an auxiliary power source. A dc-dc converter is effectively created that can transfer energy between single cells and another source of energy. Figure 14.5 shows a sample circuit concept.

The transformer could take the form of multiple secondary windings (one for each cell) and a single primary winding fed from any of the sources described above, or an individual pair of windings for each cell.

A switching circuit is required to create a time-varying current and magnetic flux for effective transfer across the transformer's core.

## 14.4 Dissipative Balancing

A lower-cost strategy employed by many battery management system designers is to simply dissipate the energy from cells that are determined to be too high in SOC using a resistive device. Although this seems intrinsically wasteful (the depleted energy is simply lost as heat and cannot be used to do any useful work), dissipative balancing systems offer a higher level of simplicity and a number of other advantages.

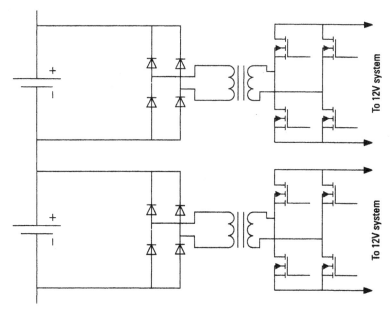

Figure 14.5    Transformer balancing concept.

**194**  A Systems Approach to Lithium-Ion Battery Management

Modern battery cells that come from high-quality manufacturers should be expected to have low self-discharge and be very similar in capacity, impedance, and self-discharge rate. This tight control of cell quality reduces the necessary balancing capability to keep a battery pack operating at rated capability.

The switches in dissipative balancing systems must only switch across a single cell voltage, minimizing the cost and size of the switch needed to turn on the balancing circuit. Despite their large number, the switches are small in size and having a single switch and resistor per cell is very inexpensive.

Many cell monitoring ICs have a provision for controlling dissipative balancing. Balancing commands are provided to the IC, which level-shifts a control signal to a voltage that is referenced to one of the terminals of the cell to be balanced. This control signal can be used to drive a transistor (usually an NMOS or NPN) to allow current to flow through the transistor and a balancing resistor. For extremely low currents, it is possible that an internal switch can control the balancing current directly, but this will limit the practical size of the battery pack that the IC can manage. Figure 14.6 shows the use of balancing transistors and resistors with a stack monitoring IC. A good compromise is an IC that allows both the use of an internal switch for low currents while allowing the same signal to drive a secondary transistor in a Darlington-type configuration for larger currents. This allows the same fundamental architecture to be

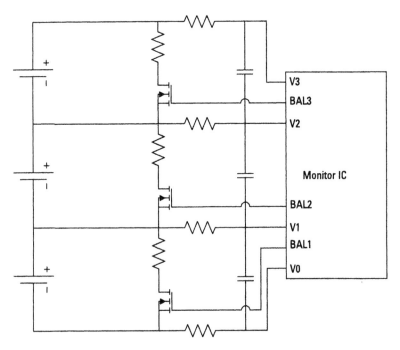

**Figure 14.6**   Typical dissipative balancing circuit.

scaled without selecting different components (which at this time often requires different control and communication architectures).

The heat generated by the balancing resistors must be appropriately dealt with. If PCB-mount resistors are used, they may likely become the warmest components on the circuit board when the balancing circuits are in operation. Only the circuits that are connected to cells that require discharging are energized, so the degree and location of heating may vary during system operation. The power rating of the components must be adequate, but more careful attention is required to the circuit board design to ensure that the overall hardware is robust. Resistors and other components may crack if they are subjected to high thermal gradients or rapid changes in temperature. This could occur with an active circuit that is adjacent to an inactive circuit. Large surface mount packages, such as 2512 components, are particularly at risk of cracking. High-power dissipation levels for large systems or high variation in self-discharge may exceed the practical limits of the surface mount package. The path by which the heat is removed from the resistors to the surroundings should be validated as well. In most cases, without external heat sinks or heat transfer devices, the path of heat flow is through the solder joints and into the PCB substrate. High heat flux can reflow or damage the solder joint. High ambient temperatures will reduce the effectiveness of nearly any cooling scheme; therefore, testing and simulation should be performed at the highest expected operational temperature of the device. Sealed enclosures generally have poor convective transfer characteristics. The use of fans will reduce the reliability of the system due to the mechanical robustness of the cooling apparatus. Because balancing resistors operate at cell stack potential, simple heat sinking to an earth-ground referenced chassis is not always a possibility. Specialty solid, gel, and foam materials that are electrically insulating can be used as a heat transfer medium but due to their cost, must be optimally selected and used. In specialty applications such as vacuum/space systems, the additional complexity of removing the heat from the balancing resistors may in fact make dissipative balancing systems more complicated than their charge-transfer counterparts.

Components in physically large packages are generally a poorer choice for dissipating heat due to larger internal stresses when heated or under mechanical load. Consider multiple smaller packages if balancing loads are high or larger-format TO263 style packages.

The reliability trade-offs with a dissipative balancing system should be carefully weighed. Increased temperature reduces the expected reliability of the neighboring electronics and can lead to higher failure rates from a system that is at a first glance expected to be simpler and more robust.

Although a single switch is adequate to control the balancing function, there are failure modes associated with failure of the switch or of its control logic. Because a closed switch leads to discharging of a cell, if the switch fails in the closed position, the cell will continue to discharge uncontrollably. Depending on the situation, this could result in the cell entering an overdischarge condition and becoming irreparably damaged. If the switch fails in the open position, balancing of that cell will be impossible. Adding switches in parallel or series will reduce the likelihood of one of these failure modes while increasing the likelihood of the other. Data about the relative frequency of various types of switch failures should be used when selecting an appropriate circuit design for dissipative balancing control.

It is usually desirable to have a self-test function for balancing to ensure that this function is operating correctly. The ultimate consequence of a balancing failure will be either slow degradation of the battery's available capacity or potential self-discharge into an overdischarged state. Either of these conditions warrants at least a basic method of error detection (if not robust and redundant protection against these events occurring).

When the battery management system operates in an active state with the battery connected to the load and real-time monitoring taking place, it is straightforward to perform cell balancing activities. The concern associated with this is that if all of the balancing must be done during the active period, and this is limited to only a few minutes or hours a day, then the balancing current during the active mode must be adequate to ensure that the required number of amp-hours per day can be discharged. This increases the cost and size of the balancing system.

A capability for sleep balancing, in which the individual cell monitoring ICs or slave devices are capable of allowing balancing current to flow during periods where the battery management system is in a powered-down or inactive state allow the balancing task to be distributed over a longer period of time. This is possible with very little parasitic power consumption as the balancing circuit can be powered by the cells directly. This allows for smaller switches and resistors but requires appropriate care to ensure that an overdischarge condition is not created. A method to control the total charge depleted during balancing must exist. This can be achieved by developing a "latching" circuit that requires an external command to change state from balancing to nonbalancing. A master device can then operate on a periodic wake-up schedule to turn off the balancing circuits when the balancing is complete. Another possibility is a timer circuit that is commanded to keep the balancing circuit closed for a prespecified period of time. This requires significantly more components but is immune to the situation where the master device never sends the stop command.

In systems in which having the battery management system in an active state leads to additional parasitic energy consumption, sleep balancing can improve overall system efficiency. Future generations of stack monitoring ICs are likely to include sleep balancing functionality.

## 14.5 Balancing Faults

Balancing errors can be divided into a few general classes: excessive discharging, inadequate balancing, incorrect balancing, or no balancing.

If a balancing switch fails to a closed circuit, a short circuit occurs elsewhere in the balancing circuit or an error exists in the software or communications path which is determining the state of the balancing switches, a cell could be excessively discharged leading to increased imbalance or ultimately causing an overdischarge condition.

It is important to note that detecting overvoltage or undervoltage conditions should disable balancing as appropriate as well as reducing overall pack current. With charge transfer balancing, it may be possible to overcharge a cell even if the contactors are opened. The same rule needs to be applied if the battery management system is no longer able to correctly measure cell voltages.

Inadequate balancing will ultimately lead to a reduction in overall energy availability. Inadequate balancing can occur for a number of reasons. If the balancing system is no longer capable of compensating for the battery's inherent variability in self-discharge and capacity, the fault is essentially a battery fault and not a battery management system failure. Battery management system defects include situations where the balancing hardware is not capable of operating and those where the software is incapable of determining and performing an effective balancing strategy.

Incorrect balancing is the result of incorrect decision-making that is caused by either a software error or by incorrect inputs. Good design practice like verifying input and output plausibility from software functions and comprehensive testing, driven by an appropriate analytical technique such as failure modes effects analysis (FMEA) is essential, considering effects as:

- Low or high cell voltage readings that lead to inappropriate balancing decisions: If a redundant measuring strategy is used or substring measurements are available, they may be used to detect inaccurate measurements and prevent them from being used to generate balancing information.

- High rates of self-discharge will necessarily degrade the state of balance over time. Tracking self-discharge rate for individual cells is beyond the allowable complexity for many battery management systems. Routine reevaulation of the balance state should be enforced to prevent self-discharge from compromising the balancing strategy. For example, imagine that all but two cells in a battery system have identical SOC, capacity, and self-discharge rate, except for two. One cell is at a higher SOC but has a higher self-discharge rate than average; the other cell has a lower SOC and a lower self-discharge rate than average. The first cell will be the one requiring the most discharging initially, but after a long standby time, the second cell will be at the highest SOC.

A secondary hardware cutoff at a prescribed low-cell voltage is a robust protection against this type of failure mode as well as software plausibility checks. Realizing this circuit using discrete components may add a significant cost per voltage measurement channel, but with ASICs this additional cost may be marginal.

# 15

# State-of-Charge Estimation Algorithms

## 15.1 Overview

Estimating a battery's state of charge provides a fuel gauge to the load device that indicates amount of charge (coulombs), which currently can be discharged from the battery divided by the total charge of the battery in its fully charged state. For the load device, it is useful to know the state of charge (SOC) to provide the user with feedback (estimated remaining run time in a backup power system, or driving range in an electric vehicle), but it is also critical for the battery management system and the battery system itself as a number of other battery parameters depend upon the SOC.

## 15.2 Challenges

Large-format lithium-ion batteries have a number of aspects that make SOC determination more difficult than with other battery types. Nonlinear relationships between SOC and OCV, which can be very flat in the case of chemistries like lithium ion phosphate (LFP) are common, and high coulombic efficiencies and the lack of self-balancing can lead to divergence in individual cells over the life of a battery system. Significant overpotentials that may have long time constants or even be hysteretic in nature often exist with strong dependencies on temperature, SOC, and age. Most significantly, the new generation of battery power applications enabled by the high energy and power capabilities of lithium-ion batteries require higher SOC estimation accuracy than most consumer electronics applications. Accuracy requirements of 3% to 5% are not uncommon for uses such as electric vehicles. Higher accuracies may well be required for aerospace or defense applications.

However, lithium-ion batteries offer a few advantages over other battery types in terms of SOC calculations. The dependence of capacity on rate (known as the Peukert effect) is much more limited than with lead-acid batteries, for example, making SOC estimation with a widely varying battery current much more feasible. Low rates of self-discharge also simplify SOC estimation.

It is important to note that SOC does not indicate the fraction of available energy in the battery. For batteries with a steep voltage profile, the lower terminal voltage indicates that the first amp-hour drawn from the battery contains significantly more energy than the last. In this case, it may be desirable to also calculate state of energy (SOE). SOE is challenging to define since the amount of energy that can be extracted from the battery depends directly on discharge rate due to the battery's internal resistance (refer to the area under the curves in Figure 15.1). However, assume that there is a terminal voltage function $V_t$ that defines the terminal voltage as a function of SOC at a given discharge rate $I_d$ (as well as temperature and other conditions). The energy function $E(SOC, I_d)$ is therefore $\int SOC \cdot C \cdot V(SOC, I_d) \, dSOC$. Depending on the application, the battery may be required to estimate SOE as well as SOC. A lookup table can usually be used to calculate SOE as a function of SOC.

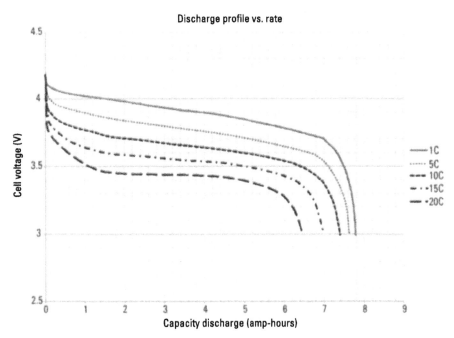

**Figure 15.1** Discharge curves for different rates.

## 15.3 Definitions

The definition of the battery's SOC is one that can vary from organization to organization and should be clearly stated and agreed upon throughout the design cycle so that confusion is eliminated.

The most straightforward definition assumes that there are two states, fully charged and fully discharged. In the fully charged or fully discharged states, the batteries are safe, stable, and not subject to excessive rates of degradation or damage. The capacity of the battery is the number of amp-hours that can be discharged taking the battery from the fully charged state to the fully discharged state or vice versa.

Ideal batteries would have the following characteristics:

- The discharge capacity of the battery is equal to the charge capacity, that is, the battery would have 100% coulombic efficiency.
- The terminal voltage of the battery is constant and therefore each amp-hour charged or discharged contains the same amount of energy in watt-hours.
- The same number of amp-hours is necessary to go from charged to discharged regardless of temperature and discharge rate.
- The fully charged and fully discharged states are not path-dependent.

In reality, these assumptions are not true, but some of them may be useful in certain circumstances. They are reviewed here:

- *Coulombic efficiency:* Lithium-ion batteries offer very high coulombic efficiencies throughout most of the charge curve; 99% or higher is not unheard of. An effective method of modeling the coulombic inefficiency of lithium-ion batteries is to use a factor $\eta$ in the relationship between the rate of change of SOC and battery current, as follows:

$$\text{During charge: } \frac{dSOC}{dt} = \eta \frac{1}{C} I \, (0 < \eta < 1)$$

$$\text{During discharge: } \frac{dSOC}{dt} = \frac{1}{C} I$$

- *Constant terminal voltage:* Due to internal resistance, OCV changes with changes in SOC, and the terminal voltage aslo includes the effects of polarization and hysteresis. Different chemistries may have terminal voltages that change significantly with depth of discharge, meaning that SOC is not necessarily an indicator of the useful energy content in the battery. For certain applications, calculating the available energy may be a separate requirement.

- *Capacity constant with temperature:* This is a poor assumption, especially over wider operating temperature ranges. Available capacity reduces with temperature in a nonlinear way.
- *Capacity constant with rate:* Lithium-ion batteries generally display better performance in this area than other types of batteries. The dependence of capacity on rate is known as the Peukert effect, which defines a power law relationship of current and capacity as:

$$C_p = I^{kp} t$$

In most lithium-ion battery systems, this effect is neglected without loss of accuracy. It is important not to mistake an earlier voltage cut off at high current with a change in the available capacity. At higher discharge rates, the magnitude of the voltage drop due to ohmic resistance, polarization, and hysteresis is greater. If the cell is discharged at constant current until the terminal voltage reaches a prescribed minimum value, the greater apparent capacity will be lower at higher discharge rates as the current is causing additional voltage depression. However, if a residual discharge is performed on these cells after relaxation occurs, it can often be shown that the remaining capacity can be discharged resulting in approximately equal capacities at different rates. At very high rates of discharge or with certain cell chemistries, it may be important to account for capacity dependence on rate (see Figure 15.1).

- *Path dependence of fully charged and fully discharged states:* The validity of this assumption varies depending on the situation. If the fully charges/discharged states are simply defined as values of the instantaneous terminal voltage, then effects of ohmic resistance, polarization, and hysteresis will create erroneous results. A common solution is to define the fully charged and discharged states as the result of a constant current/constant voltage (CC/CV) charge or discharge, which should occur at constant temperature.

## 15.4 Coulomb Counting

The simplest approach to state of charge computation is coulomb counting, or amp-hour integration. Simply put, the battery current, divided by the battery capacity, is equal to the rate of change in SOC.

This approach has a number of limitations. The first is that the method requires a correct starting point in order to track SOC accurately. It is possible in some cases to periodically obtain this starting point from a fully relaxed OCV, or from a known easily discernible voltage event, usually the end of charge or end of discharge. It may not always be convenient or possible to obtain a relaxed OCV measurement, or in some applications the end of discharge or charge is never reached and these conditions will not occur frequently. In general, it is desirable to have a method of SOC estimation that is robust against an error in the initially estimated SOC.

This method depends on accurate current measurement, especially if it is to be used over long duration cycles. The multiple-range current sensor technique discussed in Chapter 6 may improve the results of coulomb counting significantly. Consider an electric vehicle battery that is recharged over a period of 8 hours and discharged over a period of 40 minutes. The charge rate is 0.125C; the discharge rate is 1.5C. Assume a current measurement error of 0.015C, or 1% of the maximum discharge rate.

During the charge cycle, an error of 12% in SOC could accumulate, compared to only 1% during the discharge cycle. The 1% error in SOC would be considered acceptable in most applications, whereas the 12% error would not. If the error can be reduced to 1% of the actual measured value, as opposed to the maximum measured value, the error associated with amp-hour integration will be reduced dramatically.

For a battery with capacity $C$ in amp-hours, if the maximum error in current is denoted by $\varepsilon_i$, then over a period of time $t$ seconds in duration, the maximum increase in the SOC error, as calculated by amp-hour integration, is given by $\varepsilon_i t/(3{,}600 \times C)$. The time dependence of SOC error is shown in Figure 15.2.

However, coulomb counting has a number of benefits. Long integration periods minimize the effects of measurement noise on the outcome. Because error is proportional to the integration time and the current error, over short periods where currents and SOC changes are large, amp-hour integration can provide good results as the signal level of the current measurement is strong compared to the current sense errors. If the battery is regularly fully charged and discharged at a high rate (1C or higher, given the error performance of most current sensor implementations), amp-hour integration can be effective.

Battery electric vehicles and uninterruptible power supplies are examples of such applications and represent better opportunities to use coulomb-counting techniques than hybrid vehicles and peak-shaving energy storage.

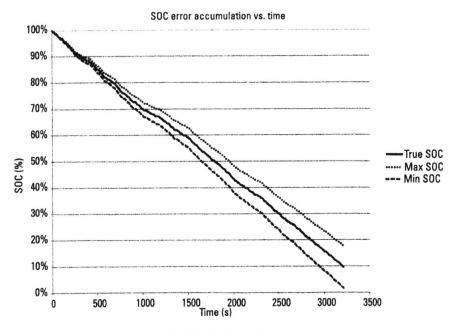

**Figure 15.2** SOC error vs. time using Coulomb counting.

In all cases, coulomb counting provides useful information about changes in SOC and is normally a part of the SOC computation algorithm for a modern battery management system. However, with any integration-based computation such as this, the issue of long-term drift will quickly become a limiting factor and the battery management system will need to rely upon other approaches to improve the long-term accuracy of the state of charge estimator.

## 15.5 SOC Corrections

It is by now clear that amp-hour integration alone is insufficient to provide acceptable SOC performance for modern battery management systems. The most basic method to provide improvement is to perform a correction of the SOC at certain defined events or points in time when the cell voltage measurements can be used to provide reliable information about the state of charge.

For applications in which the charge profile is relatively gentle and 100% SOC is reached on a routine basis. It is acceptable to perform a correction to SOC at the end of charge. With a limited charge current, the target voltage or similar method can be used to approach 100% SOC without concern for experiencing overvoltage or overcharge. When the current tapers completely, the

SOC can be reset to 100%. If the batteries are fully charged every cycle and the duration of each cycle is relatively short to prevent accumulation of integration errors, this simple strategy combined with coulomb counting may be adequate for some applications.

A similar correction can be performed at the end of discharge, with the potential concern that SOC discontinuities may be created if SOC errors have accumulated during the discharge cycle. This may not be acceptable for many applications, in which case a more complicated SOC estimation scheme will be required.

## 15.6 OCV Measurements

Assuming that a time-invariant, unique relationship between OCV and SOC exists, if a fully relaxed open-circuit voltage can be obtained, this information can be used to determine SOC directly.

For battery cells that exhibit voltage hysteresis, this method will not provide accurate results unless the hysteresis voltage is known, as there is a range of possible SOC values for a given OCV.

Complete relaxation can take anywhere from hours to days, depending on the cell. Not every application will allow the use of fully relaxed OCV measurements on a regular basis.

For cells with very flat OCV-SOC profiles, the difference of only 10–20 millivolts can represent a change of 30%–60% in SOC. Therefore, complete relaxation may be necessary to achieve acceptable error performance for SOC.

If hysteresis effects exist and cannot be accurately modeled, OCV-based estimation of SOC will be of extremely limited usefulness.

It may be possible to establish maximum and minimum possible values for SOC given the relaxation time and voltage measurements. If the maximum total polarization is known as a function of relaxation time, the maximum difference between terminal voltage and OCV can be determined, which would provide a bounded estimate of the OCV when the relaxation completes. This can, in turn, create an upper bound on the error in SOC, but for many battery chemistries (LiFePO$_4$ in particular), this approach may not provide much improvement since the discharge profile is nearly flat across a broad range of SOC, even a small hysteresis and measurement error will produce a very wide range of possible SOCs.

Such an approach may be warranted if the polarization dynamics of the cell in question are difficult to characterize accurately, the battery management system is not capable of processing a dynamic polarization model quickly enough, or the complexity of a more advanced model is not warranted.

## 15.7 Temperature Compensation

The available capacity from a battery is dependent upon temperature. Available capacity is generally reduced at low temperatures. This capacity loss is reversible upon return to higher temperatures. In the case of many applications it is important to understand the amount of charge and energy available given the current battery temperature. Therefore, the capacity and state of charge information needs to be compensated for battery temperature.

Available capacity can be modeled with a lookup table as a function of temperature. There is an important interaction between the reduction in actual available capacity and increasing impedance. The increased impedance means that less capacity is available at a given rate before the minimum allowable voltage is reached, but this energy may be extractable at lower discharge rates.

## 15.8 Kalman Filtering

Kalman filtering, developed in the early 1960s, is a mathematical technique for estimating the state (observing) of a system, designed to produce an optimal estimate of the system's state given a set of noisy input data.

Kalman filters are used in many engineering applications such as trajectory and position estimation; guidance and navigation systems rely heavily on the Kalman filter to determine best estimates of position, speed and acceleration from noisy information.

Consider a battery management system that is capable of observing current and voltage for a number of battery cells. A state space model is created that has, as hidden state variables, the battery SOC, and the state variables of the model elements. In the case of an equivalent circuit model, these could be the polarization voltages for individual RC elements and the hysteresis voltage.

In a theoretically perfect world, one single observation of the SOC is adequate to provide good SOC performance, because it is known that the SOC can be continuously calculated by integrating the battery current. Also, if the state-space model is correct and the voltage measurements are accurate, it is a simple task to extract the open-circuit voltage and therefore the state of charge from the voltage model.

However, in reality neither of these conditions is easily met. Even small errors in current and capacity lead to rapid divergence if amp-hour counting is used exclusively for state of charge measurement. The RC circuits and hysteresis models used to describe the battery behavior are only approximations of the real electrochemical effects. Furthermore, neither the voltage or current measurements are exact. Therefore, while both of these measurement and calculation techniques offer benefits to calculating battery state of charge, both are susceptible to error.

A heuristic view of these two methods suggests that they could be complementary to one another. In the short term, amp-hour integration errors will be small and from one moment to another, could provide a much more stable estimate of state of charge assuming that the starting point is accurate. However, at some point, the voltage-based calculation will be more accurate and could be used to correct or update the SOC. It is also clear that voltage errors will lead to larger SOC errors if the slope of the SOC-OCV relationship is very flat and that voltage observations may be more useful for SOC prediction at certain points of the discharge curve.

A useful analogy can be made to navigation aboard a sailboat. A vessel that started from a known point traveling at a given course and speed can determine its current position by continually plotting its course with each change of speed and direction. If the estimates of speed and heading are reasonably accurate, and other errors are small, this type of navigation (known as *dead reckoning*) provides a good estimate of the vessel's position if the time since the last known accurate position is short. As time passes, integration errors accumulate and the vessel's true position can be very far from the predicted position. Generally, dead reckoning is updated periodically using a navigation fix that gives the vessel's true position based on a different type of observation, which could be from star positions, GPS, or sighting of a landmark. The dead reckoning then continues from this new accurate location, ensuring that the vessel always has a reasonable estimate of its current position.

In the context of a battery system, the dead reckoning is analogous to amp-hour counting and the periodic fixes could be voltage-based observations. Although it is clear that using these two measurement methods together can compensate for the shortcomings of each, how should the information from each of them be combined? A number of heuristic or rule-of-thumb approaches have been suggested, but the Kalman filter solves this problem in an optimal way by computing the ideal weight given to each of the two methods.

A Kalman filter requires an underlying dynamic model for the system being observed. For a battery system, the state-space representations developed in this book can be used for this purpose. Strictly speaking, a Kalman filter requires this underlying system to be a linear time-invariant (LTI) system, which battery systems are not; the important points of this distinction will be discussed later. Kalman filters also operate on discrete-time approximations of continuous-time systems and therefore only the discrete-time version of the mathematics will be shown.

The Kalman filter assumes that both the system model and the observation contain noise. The *process noise* influences the state evolution of the system and accounts for errors in the inputs as well as the model of the system's true dynamics.

$$X_k = AX_{k-1} + B_k u_k$$

With process noise $w_k$

$$X_k = AX_{k-1} + B_k u_k + w_k$$

The *observation noise* models the errors in the measurement of the system's outputs.

The output of an ideal system $Z_k$ are given as a function of the state variables.

$$Z_k = C_k X_k$$

In a nonideal system, the output is perturbed by the observation noise, $v_k$

$$Z_k = C_k X_k + v_k$$

Assuming an accurate dynamic system model and measurements of inputs, the new state of the system at the new time instant can be predicted. For a battery system, the input to the new model is the battery current and the output is the terminal voltage. It is logical to think that if the battery model is correct, the state of charge, the terminal voltage and the voltage of each polarization and hysteresis element can be predicted at the end of the time step.

If the model and measurements are fully accurate, then the vector ($\hat{z}_k - z_k$), known as the *measurement residual*, comparing the predicted measurements with the actual measurements will be zero. In an imperfect system, this vector will vary in magnitude depending on the degree of error in the measurement and the prediction.

The Kalman filter, at each time step, computes the *Kalman gain* $K_k$, which is the blending factor used to optimally weight the predicted state of the system from the model dynamics and system inputs, and the corrected state based upon the measured outputs of the system to generate an optimal estimate of the system state.

It is assumed for the true Kalman filter that the process and observation noise are zero-mean, Gaussian, and white, with a known covariance of $R_k$. It can be shown that even in systems where these assumptions about the noise do not strictly hold, that the Kalman filter provides excellent performance. However, it cannot be shown that the weighting calculated by the filter is ideal.

The process and measurement noise have covariance matrices **Q** and **R**, respectively. The diagonal values of **Q** and **R** represent the variances (square of the standard deviation) in the process and measurement variables, and the off-diagonal elements represent the covariance between state variables. The determination of the values for **Q** and **R** depend on the accuracy of both the measurements and the model. The measurement accuracy for quantities like voltage and current can be determined through testing and assumption that measurements are zero-mean and Gaussian is likely valid for many types of sensors (magnetic Hall effect current sensors may be a notable exception if a bias exists). In many cases it is also likely that the measurement covariances will be zero—that there will be no correlation between noise in voltage measurement and current measurement, for example. The process errors will be harder to quantify.

The matrix **P**, which is the a posteriori state error covariance matrix, is updated with each time iteration of the Kalman filter. This matrix specifies the degree of uncertainty associated with each of the state variables. The **P** matrix can be used to determine specifically the degree of trust in the state of charge estimate that is calculated. The selection of the initial values for **P** specifies how much the initial state variables are trusted. In the context of a battery management system, consider that the initial state could be the state of charge and overpotentials loaded from nonvolatile memory at power-up of the battery management system. This data could be from a recent operation cycle that has recently calculated the SOC accurately, or it could be from an operation cycle in the distant past where overpotentials have relaxed and self-discharge has depleted the SOC. In the worst case, the battery management system could be powered up for the first time with the battery system and have no prior knowledge about the battery state. The covariance matrix should be set with an appropriate initial value depending on the degree of confidence in the initial state variables.

The initial covariance estimate can be manipulated depending on the degree of certainty in the previous state of charge estimate. Depending on the degree of self-discharge expected, the length of time that the battery system has not been monitored by the battery management system, or for brand-new battery cells in an unknown condition, it may be sensible to set the covariance estimate small if the previous SOC estimate is thought to be highly reliable, or large, if there is relatively little useful information about the SOC of the battery.

The process noise covariance matrix **Q** is also complex to determine. In most situations, the model used will be an approximation of the true battery dynamics and therefore the errors are not strictly random in nature; the deviations in the implemented model are deterministic, but their effects may be pseudo-random depending on the circumstances. For example, consider an equivalent circuit model where a constant phase element is replaced by a capacitor; in many cases, this is a good approximation and simplifies the battery model considerably. The state variable associated with the capacitor is the voltage across the element.

The response of the capacitor differs from the constant phase element (which is in itself an approximation of the actual physics within the battery). The magnitude and phase of the error depend on the frequency of the excitation, and therefore it is impossible to say that the error component caused by this approximation is Gaussian.

Despite the differences between actual battery systems and Kalman filter theory, under a wide variety of conditions, this method has been shown to produce robust and reliable SOC estimation that is immune to errors in initial estimation, resistant to integration error as well as model inaccuracies and measurement noise. Numerous academic and commercial examples of this technique exist [1].

For nonlinear systems, an extension to the Kalman filter called the extended Kalman filter (EKF) exists. The state transition and observation matrices are replaced with nonlinear functions.

Nonlinear functions can capture specifically the nonlinear relationship between SOC and OCV. The partial derivative of OCV with respect to SOC is the slope of the SOC-OCV curve and allows an EKF-based SOC estimator to automatically incorporate the changing relationship between voltage measurement error and SOC error depending on SOC.

Other nonlinearities to be captured may include different ohmic resistances in charge and discharge, nonexponential modeling of hysteresis, and more advanced representations of equivalent circuit elements such as the CPE and Warburg impedance.

The EKF requires computation of the Jacobian, a matrix of partial derivatives. Calculating this online, numerically, or analytically is often computationally expensive. However, because many of the nonlinear functions represented use lookup tables and are constant (or approximately so) throughout battery operation, the partial derivatives can be calculated and stored along with the lookup table values. The derivative of the SOC versus OCV relationship is required, as well as the derivative of model parameters with respect to SOC. These are relatively easily handled by lookup tables. More importantly, because the uncertainty in the system is propagated through the linearization, the propagation of the process and measurement noise may introduce additional errors, depending on the degree of uncertainty in the system. Figure 15.4 indicates how this may occur.

Another derived technique is the unscented Kalman filter (UKF), which avoids the calculation of the Jacobian. The UKF performs error propagation through selecting a number of points around the current state estimate and transforming them through the nonlinear functions to obtain a more realistic estimate of the new distribution.

A comparison of the linearization techniques of an EKF and UKF is shown in Figure 15.3.

An important benefit of the Kalman filter for SOC and battery state prediction is that the method requires only the values of the state variables from the previous time step. An extended history is not required, minimizing memory requirements in a real-time embedded system.

## 15.9 Other Observer Methods

Many other methods for observing the states of a nonlinear system exist and have been proposed for SOC estimation, including high-gain observers and sliding mode observers. Other efforts have focused on the use of fuzzy logic or artificial neural networks (ANN) for SOC estimation. These novel methods may gain relevance as research continues into SOC estimation techniques.

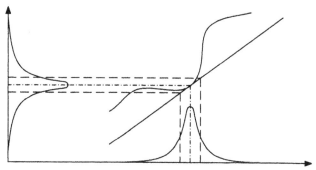

EKF uses Jacobian linearization creating incorrect probability distribution

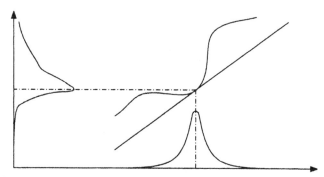

UKF samples probability distribution of independent variable for more accurate distribution of dependent variable, but requires more computational effort.

**Figure 15.3** UKF and EKF linearization realization.

## Reference

[1] Plett, G., "Extended Kalman Filtering for Battery Management Systems of LiPB-Based HEV Battery Packs," *Journal of Power Sources*, Vol. 134, 2004, pp. 252–261.

# 16
# State-of-Health Estimation Algorithms

## 16.1 State of Health

The concept of a battery's state of health is an abstract concept that attempts to reduce the complex phenomena that combine together to produce battery degradation to a simple metric indicating how far the battery has progressed from beginning of life to end of life. The definition of end of life varies in individual applications and may have multiple possible definitions, but in general, when a battery system is no longer capable of providing the minimum power, energy, and standby time needed for the application, the battery system is in need of servicing or replacement. A number of internal processes lead to three significant externally observable effects which constitute reduction in battery health, namely capacity fade, impedance growth and increased self-discharge.

Battery capability fades as a function of the number of charge/discharge cycles (usually known as *cycle life*) as well as total time in service, known as *calendar life*.

*Capacity fade* is a reduction in the available energy and charge capacity of the battery over time. Internally, capacity reduction occurs due to one of two root causes: inability of either lithium ions or electrons to reach active material sites. These problems can be due to a number of other effects including damage to the electrode structure either at the microscopic level or macroscopic level (active material may dislodge from the electrode surface and no longer be in contact with the current collectors). A common industry criterion for end-of-life capacity is considered to be 80% of initial capacity, but this may be significantly more or less than what is required for a specific application.

Impedance growth leads to a reduction in the rate capability of the battery. Many of the same phenomena that cause capacity fade also contribute to impedance growth. Most lithium-ion batteries using carbon anodes experience soild-electrolyte interphase (SEI) growth that increases the impedance during battery aging. The loss of active material results in the surface area for reactions being reduced and therefore higher impedance. Electrolyte degradation and increases in interface resistance also contribute to increases in impedance. Assuming that the limit voltage is fixed throughout aging (the validity of this assumption may vary), then the allowable charge and discharge rates that will lead to reaching this limit voltage will reduce correspondingly. Allowable power fade varies more widely than capacity fade with some applications tolerating as much as a 50% reduction in available power.

Self-discharge rates may increase as lithium-ion batteries age. As the self-discharge rate increases, the available standby time of the battery is reduced. Assuming that the battery management system is sized for a nominal beginning-of-life self-discharge profile, increasing and diverging self-discharge rates from cell to cell will reduce the battery management system's ability to compensate, and eventually will reduce battery performance.

These three factors are used to compute a single metric, the state of health (SOH) of the battery systems.

$$SOH = SOH\left(Capacity, Impedance, Self\ Discharge\right)$$

An idealized SOH value is between 1 and 0 and is commonly expressed as a percentage. When the battery is new, the SOH should be 100%. As soon as the battery reaches the point where it is marginally able to provide the power, capacity and standby time that are needed for the application, the SOH is usually defined as zero. Optimally, if the battery is cycled using an identical charge/discharge profile under identical environmental conditions (such that each cycle is effectively the same), then the SOH should reduce in a linear fashion relative to total cycle count. In reality, operation at more extreme rates, temperatures, and states of charge will, in general, reduce the health of the battery faster than under mild conditions and the reported SOH will not decrease uniformly under different types of cycles.

An accurate SOH estimation system therefore requires a clear definition of the beginning-of-life and end-of-life limits for the cell's capacity, impedance and self-discharge rates. An ideal battery management system must be capable of estimating these three parameters dynamically during the operation of the battery system using only the same inputs of voltage, temperature, and current that are available. In laboratory testing, a *reference performance test* (RPT) is performed throughout cycle and calendar life testing to evaluate the capacity and impedance of the batteries. Some applications may permit some form of performance test to be performed on the battery cells, but, in general, the battery management system will have to analyze data gathered from normal battery operation to determine SOH. A combination of predictive factors including measuring time in service and total amp-hour throughput, and reactive factors including online capacity and impedance estimation is used to generate an optimal estimate of the remaining useful life of the battery.

The estimation of SOH is further complicated by the fact that the capacity, impedance, and self-discharge are affected by temperature and other transient factors. These should not influence the indicated SOH, which should only reduce with battery use. The SOH estimator is expected to determine the battery's capability relative to a new battery at some nominal condition, despite the fact that the instantaneous battery capability will vary considerably. SOH calculations should consider the relatively slow decline of battery performance over hundreds or thousands of cycles, but still offer enough responsiveness that a defective battery that declines quickly can be detected. The problem of SOH and remaining useful life estimation can be described as nonexact (battery models do not completely model the performance of the cell), nonlinear (battery characteristics such as SOC-OCV relationships are highly nonlinear), and nonstationary (battery parameters change with SOC, temperature, and time) models with non-Gaussian noise [1].

## 16.2 Mechanisms of Failure

The end symptoms of capacity loss and impedance increase are the externally observable effects of many complex interactions occurring within the battery. A basic understanding of how a lithium-ion battery loses capacity is helpful in the development of battery management systems for this type of battery systems.

During charging, a lithium atom, intercalated in the active cathode material must de-intercalate and become oxidized, losing an electron. The electron must travel from the active material site, through the positive electrode, ultimately to the current collector and to the positive terminal of the battery. The lithium ion must travel from the active material site, through the electrode material, into the electrolyte, where it must cross the separator to reach the anode. The ion must reach a particle of active anode material and recombine with an electron that has made its way to the same place, via the negative current collector and terminal, and reintercalate into the negative electrode material. The process proceeds in reverse during discharging. Considering the steps involved, capacity fade and impedance growth can be interpreted as failures or impediments in one of the steps of the above process.

Loss of active cyclable lithium ions leads to a reduction in capacity. This can happen in a number of ways. The passivating film on the anode layer, known as the solid-electrolyte interphase (SEI) is produced by the reaction of lithium ions and electrolyte. The exact reactions and products are poorly understood, but lithium is consumed during the formation of the SEI layer, which occurs significantly during the first few formation cycles of the battery but continues at a slower rate thereafter. The most described effect is the reduction of solvents in the electrolyte to form lithium-containing compounds. The SEI growth reactions are generally seen to occur primarily during charging. The formation of the SEI is responsible for the large portion of the capacity loss during the initial period of the battery's life. Lithium ions can also be lost due to plating of metallic lithium due to the anode voltage falling below zero with respect to Li/Li$^+$, most often the result of charging too quickly at low temperature, but this is not an expected normal method of capacity loss in a properly maintained battery. Lithium may also be lost through other, less significant side reactions such as electrolyte decomposition consuming the lithium salt.

The other possibility for capacity loss is loss of active material; without sites for intercalation, the charge transfer cannot occur. The active material can be degraded through side reactions with the electrolyte. The most common hypothesis is that the active material becomes isolated from either the electrolyte, preventing access by lithium ions, or isolated from the current collector, preventing access by electrons. Either of these causes the loss of available sites for intercalation and reduces the battery capacity. Prevention and root causes of these effects are still a subject of active research; the volume change of electrode materials during charge and discharge is commonly accepted as a major cause of breakdown. If active material is lost in the lithiated state, lithium ions are lost as well.

Impedance growth is caused by resistance to the transport of electrons and ions in the various components of the battery. The ohmic resistance of the metal current collectors, tabs and terminals is expected to be relatively stable and not a source of significant resistance changes. The growth of the SEI layer adds resistance to the transport of lithium ions into and out of the anode.

The above reactions are certainly sensitive to temperature (higher temperatures lead to faster degradation) and current, specifically charging rate, as well as the final charge and discharge limits selected. Models have been created to account for the effects of all of these inputs on degradation rate.

## 16.3 Predictive SOH Models

It is possible to gain an understanding of the expected rates of capacity decline and capacity fade through extensive testing of battery cells under controlled conditions. Many researchers have attempted to create first principles models of the degradation mechanisms outlined in the previous section with varying degrees of success.

However, the computational complexity of these models usually is such that they cannot operate in real time, and often the models do not necessarily reproduce all observed effects with high fidelity. First-principles models are often replaced with models constructed from empirical observations of cell behavior during cycling and calendar life testing.

These models attempt to model the capacity and impedance of the battery as a function of the following independent variables:

- Temperature exposure;
- Calendar life;
- Depth of discharge;
- Cycle life or amp-hour throughput;
- Battery current.

In some applications the cycle profile is relatively consistent in terms of discharge and charge current, as well as depth of discharge for each cycle, and the temperature does not vary significantly. In these cases, it is convenient to look at the capacity fade function as a function only of the number of cycles.

Typically an approximation function is used to model this capacity decline. The use of an appropriate approximation function is critical and depends on appropriate use of analytical and experimental techniques.

Alternatively, it may be helpful to look at a discrete-time differential representation of this relationship. The capacity of the battery during cycle $C(k)$ can be expressed as a function of

- Capacity during the previous cycle $C(k-1)$;
- Temperature during cycle $k-1$;
- Charge and discharge currents during cycle $k-1$;
- SOC window during cycle $k-1$;
- Time duration of cycle $k-1$;
- Initial capacity $C_0$.
- Total time in service ($t$).

$$C(k) = f\begin{pmatrix} C(k-1), T(k-1), I(k-1), \\ SOC_{max}(k-1), SOC_{min}(k-1), t(k)-t(k-1), C_0, t(k) \end{pmatrix}$$

The initial capacity and total time in service terms are required because the rate of various degradation mechanisms changes over the life of the battery. For example, the SEI growth function tends to slow over time; therefore, the capacity loss due to this effect between cycles 10 and 11 will be more significant than between cycles 100 and 101, for example. See Figure 16.1 for typical degradation curves for lithium-ion batteries.

The influence of these parameters should be determined through a careful design of experiments to separate the effect of each parameter on life. In most cases, it will be determined that, based on the cell performance and the parameters of the application, a number of the parameters do not need to be represented in the online model that will be implemented in the battery management system. For example, in a hybrid-electric vehicle the window of SOC used in operation will be very small and there is not likely to be a significant difference between the degradation rate in the most charged and the most discharged states of the battery. In cells in which the SEI layer growth is the dominant effect driving capacity loss, the effect of varying the discharge current may be largely irrelevant compared to the charge currents. Because, in most cases, a corrective element to the SOH model will be established, the predictive component need not capture all secondary effects. The simulation of models of varying complexity to test their effectiveness is a useful method to select the dimensions of the parameter space for a degradation model.

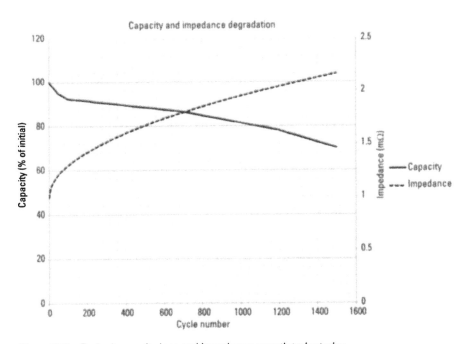

Figure 16.1 Typical capacity loss and impedance growth trajectories.

Once the relevant parameters that impact cycle life are determined, the curve should be established to fit the experimentally gathered data. This can be done through the use of appropriate basis functions or by using look-up tables.

Power-series and exponential basis functions as well as polynomials have been used to fit degradation data with varying degrees of success.

These basis functions are used to predict the path of the battery performance curve as a function of variables that influence SOH. Battery testing using well-designed experiments and careful analysis of the root causes of degradation can determine the coefficients of these functions appropriately and reduce the parameter set to those which influence the SOH meaningfully.

## 16.4 Impedance Detection

As impedance growth is one of the primary symptoms of lithium-ion battery degradation, measuring battery impedance is an important fundamental capability.

Methods for determining battery impedance can be classified into two broad groups: active methods, in which current is injected into the battery for the sole purpose of making an impedance measurement, or passive methods, in which only observations of the current requested by the load can be used.

### 16.4.1 Passive Methods

An applied step function in current and the resulting voltage profile is shown below. Upon the initial application of current, there is a corresponding rapid change in voltage corresponding to the ohmic resistance of the cell, followed by the buildup of overpotentials associated with diffusion and reaction kinetics.

A reasonably accurate estimation of the dc ohmic impedance can be obtained using a linear least-squares fit (see Figure 16.2) to a series of voltage and current ordered pairs. The slope of the regression line is equal to the impedance and the intercept of the line is referred to as the IR-free voltage. This method is relatively robust against measurement noise in voltage and current. It is more accurate in cases where the current profile is roughly equal in charge and discharge both in terms of total amp-hours and size of the peaks (in this way the polarization dynamics can partially cancel each other). It is also more accurate with shorter pulses in which polarization has less time to develop.

The ratio of $\Delta V/\Delta I$ can be used to calculate ohmic resistance, either at discrete points or on a continuous basis.

The discrete approach detects reasonable approximations of step changes in current and measures the corresponding instantaneous change in voltage (see Figure 16.3). This method assumes that step changes in current occur frequently and over a wide range of operating conditions. This may not be true

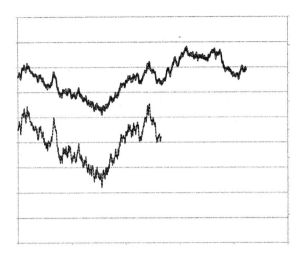

(a)

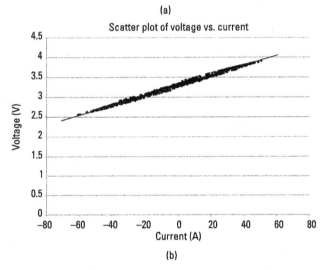

(b)

**Figure 16.2** Linear V-I plot of battery response.

for all load devices. This method is more accurate with large changes in current and therefore a minimum step size may be used or a weighting function giving greater weight to large changes. Like most impedance measurement methods, time synchronization between the voltage and current measurements is critical so that the ohmic effects are fully captured and other voltage changes are neglected.

Changes in voltage and current can also be calculated on a continuous basis. In the case of steady current flow while voltage changes as the cell polarizes, the result of this becomes nonsensical (because $\Delta I$ is zero). Therefore,

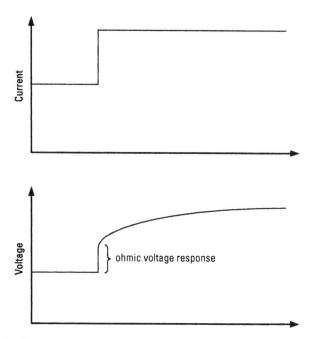

**Figure 16.3** Ohmic voltage drop response.

this special case needs to be ignored. There may be better accuracy because the calculation can be performed more regularly than with the discrete method; however, it is more difficult to disregard the effects of cell polarization.

### 16.4.2 Active Methods

The above methods can be enhanced by injecting controlled current profiles. For systems that are periodically connected to a battery charger or to an appropriate load device, it may be possible to inject current profiles, either ac or dc, of known magnitude, and synchronously perform voltage measurements and extract more accurate impedance data.

For a given cell at a given point in its aging cycle, the ohmic resistance is a function of SOC and temperature. As such, a broad set measurements cannot be simply all averaged together to obtain a true picture of the ohmic resistance. It may be desirable to again use an opportunity calculation and only perform impedance measurements at certain points in time. Another option is to perform impedance calculations continuously while state of charge and temperature are known, and use the SOC and temperature data to update a particular value of the impedance function at those values of SOC and temperature.

For a given SOC and temperature, impedance changes take place slowly. A typical rate of impedance increase might be 0.01%–0.1% per cycle. Due to the inaccuracies in all the calculation methods, it is therefore desirable to introduce long-term lowpass filtering into the impedance estimation.

If online model parameter estimation is used, the nonlinear least squares method and the outer Kalman method will provide impedance estimation.

For batteries with large differences in ohmic resistance between charge and discharge, these above methods can all be extended to determine the charge and discharge resistances separately.

Many battery systems use interconnects and busbars of different size and geometry between different groups of cells. These busbars may correspondingly have different impedances. As such, the varying busbar impedances may give the appearance of a wide variation in impedance. It is common practice to compensate for the effects of busbar impedance based on the position in the cell stack, using bus bar impedance values either measured or calculated to give the best estimate of the battery cell only. This positional compensation should be used for battery model operation as well to remove the effects of busbars on SOC calculation and other higher-level operations. Care should be taken in the layout of the battery pack and the battery management system interconnect to ensure that if possible any potentially high impedance components (fuses, service disconnects, and so forth) are installed in such a way to ensure that the battery management system is not measuring the potential drop across these components. See Figure 16.4 for an illustration of this.

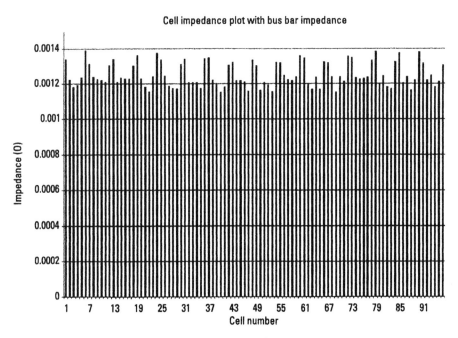

Figure 16.4 Typical impedance plot showing effect of busbars.

Additionally, battery impedance and busbar impedance will be temperature-dependent. For metals, impedance increases as temperature increases but battery impedance generally shows a negative temperature dependency. Busbar compensation values may need to be temperature-dependent and battery impedance is then measured at a given temperature. Like the capacity-temperature relationship, an impedance-temperature-SOC relationship over the entire operating range must be built from imperfect data gathered from observing normal battery cycling and no general relationship is known that will explain the increase in impedance at one temperature associated with the increase in impedance measured at a different temperature.

Once an accurate measure of cell impedance is available, it is important to understand how this information should be used to predict state of health. One should start by defining the power capabilities in charge and discharge of a marginally acceptable battery (i.e., a battery that just meets the requirements of the application). This should specify a certain charge and discharge power capability, over a given SOC window, temperature range, for a specific duration. Given a model of the cell's polarization and hysteresis, the resulting dc impedance of this marginal battery cell can be determined. This effectively defines a set of 0% SOH conditions (there may be more than one, because the battery may have different impedances at different SOCs and temperatures and in charge and discharge).

Usually the growth of impedance in battery cells is not linear with the number of cycles or time in service. Increased impedance is due to a number of proposed mechanisms, including growth of the SEI layer, degradation of the electrodes at different levels of scale, side reactions between the electrolyte and electrodes, and others. Many models of impedance growth specify a behavior that is proportional to the square root of the time in service. The battery therefore shows a higher rate of impedance growth in the beginning of life than later. If this nonlinearity is significant, the SOH may decline rapidly in the initial cycles and this may provide misleading information to the user. Therefore, it could be desirable to linearize the SOH using a generalized impedance growth curve to make a first-order approximation of how many cycles remain before the impedance maximum is reached.

## 16.5 Capacity Estimation

In addition to short-term variation due to temperature, the battery's capacity is not fixed over the life of the system. The battery service life is usually long enough that the capacity is expected to degrade but the overall battery system should remain useful.

Reasons for capacity decline are discussed above. It may make sense to separate individual causes of capacity decline if the battery operating parameter space is very wide and the shape of the degradation curve is expected to be different across all possible applications.

The capacity needs to be measured dynamically for two reasons: first, to provide the user with useful feedback about how much energy remains in the battery system, and second, to help the system and user determine when the capacity has degraded to the point that the battery cells must be replaced.

It is important that the capacity estimation provide a realistic estimate of actual long-term reduction in capacity without being distorted by short-term variations in available capacity due to temperature and discharge rate.

In systems that routinely experience full discharge and charge cycles, capacity estimation is the most straightforward. With each cycle, an actual capacity measurement can be made and a lowpass filter applied to smooth out variation in capacity measurement. Due to temperature influence on capacity, this strategy works best if the battery operates at a near-constant temperature, or if an effective temperature compensation relationship (which, in the best case, is approximately invariant as the battery ages) can be created. Some applications are fortunate enough to have this type of duty cycle but in many cases the charge and discharge levels vary from one cycle to another.

Conceptually, the process of capacity estimation with an online battery model works as follows:

- During individual discharge cycles, the battery state is estimated using a predictor-corrector approach. The model predicts the battery state at a future time and the model is refined using corrections if the predicted battery values do not match measurements. Normally this is realized using the integral of the current to predict the change in state of charge and voltage measurements to produce corrections. Many ways of doing this are discussed.

- If the deviations from the predicted values occur solely due to random noise in both the model and in the measurements, it should stand to reason that there should be no long-term bias in the sense of the correction (i.e., both state of charge overestimation and underestimation should be equally likely).

- If however a meaningful bias is detected over the course of multiple cycles, there is an increasing probability that this repeated error is caused by an error in the capacity value. For example, if the battery voltage during discharge is consistently predicted lower than expected (assuming that an accurate model accounting for all other effects correctly), it is likely that the capacity parameter is incorrect and the battery is simply more discharged (and therefore at a lower voltage) than predicted.

- These long-term bias estimates can be used to generate an update to the battery's capacity. It is normal to assume that a lowpass filter with a very long time constant must be applied to this data due to the slow change in battery capacity over hundreds or thousands of cycles. Long-term filtering will effectively minimize the effect of random noise and ensure that erroneous results do not create rapid changes in battery capacity.

Because available battery capacity is a function of temperature, a significant challenge arises around producing a reliable capacity estimate. A capacity function can be established which describes the available capacity as a function of temperature, usually describing the available capacity as a fraction of the capacity available at some reference temperature (usually 25°C). In a laboratory setting it is possible to hold the cell at the reference temperature to reestablish the reference capacity as a cell ages and to determine the capacity function. Performing this type of measurement in-service is not possible in many applications. Therefore, capacity estimates may require incorporating information measured at a variety of temperatures other than the reference temperature. Furthermore, this range of historical temperatures may not cover the total operating temperature range of the battery, implying that the battery may one day be exposed to a new temperature extreme where a capacity estimate has never been performed. Lastly, the distribution of temperature exposure will not in general be uniform meaning that most data will come from a narrow range around the mean temperature and relatively few measurements will exist at extreme temperatures. A battery management system must create a usable capacity estimate despite all of these difficulties.

If an effective capacity degradation model exists, it can be placed inside the battery management system as a feedforward predictor of the capacity decline. A state-observer system, such as a Kalman filter, can be used to generate an estimate of the battery capacity in this way. This will help limit the effects of noisy measurements of capacity by using a proven laboratory aging model as a baseline degradation trajectory.

A minimum effective capacity should be established at which the battery is marginally acceptable at which the SOH should be considered 0%. A new battery with rated capacity or more would have an SOH of 100%. As the rate of capacity decline is usually not linear, a time linearization can be performed to ensure that the SOH declines in a linear fashion with time and not with battery capacity.

## 16.6 Self-Discharge Detection

The final factor affecting state of balance is excessive increases in self-discharge rate. If the battery is unable to retain charge to meet standby requirements or the self-discharge of individual cells exceeds the ability of the battery management system to correct the imbalance, capacity loss will usually occur swiftly from the divergence of the cells.

Self-discharge may increase due to cell defects (including dendrites of metallic lithium or other conductive particles) that breach the separator, allowing current to flow from the positive to the negative electrode inside the cell.

If effective balancing strategies are in place, and the rate of self-discharge is lower than the allowable balancing current, then the cells should remain in balance, and the duty of the balancing should be less than 100%. Increased differences in self-discharge will lead to larger balancing duty cycles.

## 16.7 Parameter Estimation

If the same methods for determining battery model parameters are run in real time within the battery management system software, and the battery model parameters are updated over the life of the battery system, the changing values of the parameters can be used as an effective measure of battery SOH. Refer to Chapter 12 for details.

## 16.8 Dual-Loop System

A battery model assumes that the battery parameters are constant during battery operation and attempts to estimate battery state of charge and other internal state variables. In the same way that a battery model attempts to predict the voltage of the battery cells under a given current profile, which is then updated by a measurement of the actual voltage, a similar predictor-corrector loop can be run on the model parameters themselves. It is necessary to assume that the battery model parameters change necessarily much more slowly than the internal states for this strategy to work; in a given cycle, the capacity and impedance are expected to be roughly invariant.

An example of such a system (shown in Figure 16.5) uses a Kalman filter for both SOC estimation and parameter estimation. The SOC estimator uses voltage, current, and temperature measurements to update the internal states of a battery model used to predict SOC. At the same time, the parameter estimator uses the same information to calculate new parameter values. The concept is straightforward (consider that the SOC estimate is always too high); it is conceivable that the capacity of the battery has decreased.

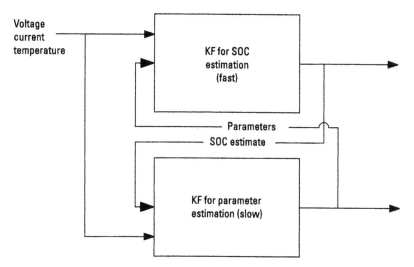

**Figure 16.5** Dual-loop extended Kalman filtering block diagram.

## 16.9 Remaining Useful Life Estimation

Remaining useful life is defined as the expected remaining time for which the system can be expected to meet its intended function. Whereas SOH computation looks only at the level of degradation that has occurred with respect to the maximum allowable degradation for a system considered marginally functional, remaining useful life prediction must also make some estimates about current and future rates of degradation.

## 16.10 Particle Filters

Particle filters are a class of filters intended for observation of state space systems with highly nonlinear dynamics. They provide good performance when the system noise is non-Gaussian. They have been used for capacity fade and impedance growth models that are challenged by especially noisy measurements and highly nonlinear behavior.

A particle filter, similar to other state observers, is useful for performing regression analysis to fit measured capacity and/or resistance data to a predicted degradation curve.

As with most SOH approaches, this method requires a predictive method of predicting the SOH variables at cycle $k$ given the values at cycle $k - 1$.

Additionally, a method of estimating the SOH at each cycle is required. It is necessarily assumed that this method will contain errors in the estimation of the SOH and that the estimation of the SOH is distributed with some probability distribution function around the true SOH.

A series of particles, or possible estimates of the true capacity, is generated randomly. Given these theoretical capacities, the measured capacity, and the probability distribution (which need not be Gaussian or of any other type), the probability that each of the particles is correct, given the observed capacity, can be calculated using the Bayes theorem.

The particles are then transformed via the predictive function to a new expected SOH at a new cycle. For the $i$th particle estimating capacity $Ci$ at cycle $k$, a new capacity $Ci_{k+1}$ is generated using the update function. See Figure 16.6 for an example of the results.

The set of particles is then resampled to select only those particles that meet a minimum probability of being the true capacity. This leads to the particle filter's other name, sequential importance resampling.

Rather than a closed-form function for the probability distribution, the distribution of the capacity estimates is represented by the collection of particles and their respective probabilities.

The particle filter technique provides an estimate of both the state of health and the accuracy of the estimate.

Particle filters have been used successfully for SOH prediction with good robustness to nonlinearities and very noisy calculations of capacity and impedance. A useful degradation model is still an essential element to achieve good performance.

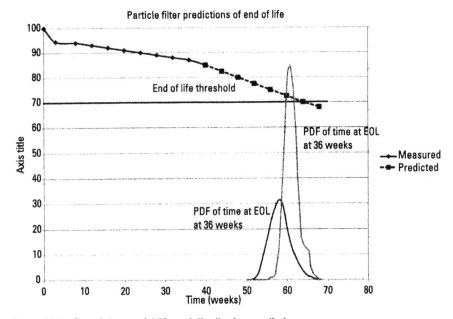

**Figure 16.6** Remaining useful life and distribution prediction.

# Reference

[1] Saha, B., and K. Goebel, "Modeling Li-ion Battery Capacity Depletion in a Particle Filtering Framework," *Proceedings of the Annual Conference of the Prognostics and Health Managemetn Society*, San Diego, CA, 2009.

# 17

# Fault Detection

## 17.1 Overview

The battery management system is required to detect defects in the battery cells as well as the balance of plant to ensure that a damaged battery that is susceptible to becoming hazardous with further use is not placed in service. These detection mechanisms must be accurate but robust against nuisance errors which reduce battery reliability and availability. Reactions to battery failures must be timely and appropriate and coordinated between the battery management system and the load devices to preserve safety.

Failure rates for lithium-ion cells depend on the maturity of the manufacturing processes for the particular cell. The 18650 cylindrical cells are manufactured in extremely high numbers and failure rates well below one internal defect per million cells are achieved by most reputable manufacturers. However, when these cells are used in large arrays, the failure rates at the system level may become appreciable again. Other cell technologies may be significantly less mature and have a much higher probability of a cell defect.

## 17.2 Failure Detection

### 17.2.1 Overcharge/Overvoltage

Detection of overcharge or overdischarge is crucial to avoiding the multiple hazardous effects that these conditions create.

The first line of defense is accurate and complete cell voltage measurements. Faults should be set at the edge of the normal operating range and the system should respond rapidly by reducing limits and disconnecting the battery if overcharging or overdischarging is occurring.

A second layer of protection can be incorporated by measuring the rate of change of voltage (see Figure 17.1). As most cells approach overcharge or overdischarge, the cell voltages begin to rise or fall much more rapidly than in normal operation.

Bounds can be set on the maximum expected rate of voltage change with respect to cell current. The slope of the OCV(SOC) curve is the rate of change of voltage with respect to rate of change of state of charge. Normalizing with the battery capacity gives the following.

$$dV/dSOC = C\, dV/dQ$$

This can be converted to a time rate of change by multiplying by the rate of change of charge ($dQ/dt$), which is the battery current

$$dV/dt = C\, dV/dQ\, dQ/dT = C\, dV/dQ\, I$$

When rapid changes in battery current are possible, corresponding high rates of change in cell voltages are possible, which will need to be compensated for.

From this, a maximum expected rate of change can be established for the IR-compensated voltage (or, with more effort, the open-circuit voltage) as a function of the SOC and current. If these rates of change are exceeded,

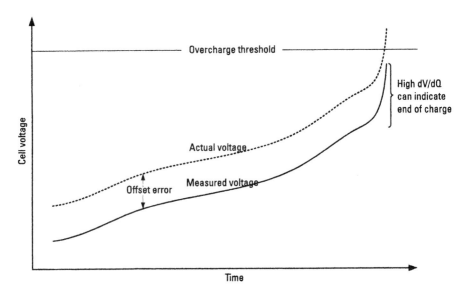

Figure 17.1  End-of-charge voltage error detection scheme.

it is possible that the battery is experiencing one of the following hazardous conditions.

- The cell is exceeding the safe voltage range and entering overcharge or overdischarge, but a voltage offset measurement error is causing the cell measurement to appear safe.
- The cell is approaching the edge of the safe operating area. For systems with high power-energy ratios that operate over wide SOC windows, this technique can be used to allow higher charge and discharge rates without risk of overcharge, overdischarge, or cell reversal. This could be due to errors in SOC estimation.

An additional layer of protection can be created using backup strategies to calculate the total accumulated charge added or removed from the battery. For example, a fully discharged 50-Ah battery should theoretically not be able to accept more than 50 Ah without becoming overcharged. If the charge process is cut off after 60 Ah are added, this can limit the overcharge to 20%. Straightforward coulomb counting may not provide accurate SOC performance, but it may prevent or limit overcharging in the event of the failure of other means of protection.

For battery systems that are charged at high rates to high states of charge, overcharge may happen very quickly. Most lithium-ion cells will display a rapid increase in voltage once the cell exceeds 100% SOC. An effective means of preventing sudden overcharge, if the battery charger can operate with a voltage limit, is to continually command the charger to enforce a pack voltage limit which is only slightly higher than the current battery pack voltage, and to continually update this limit, "floating" it just above the measured pack voltage. If an individual cell voltage rises suddenly, the charger will encounter the limit voltage and stop charging. A full pack voltage limit at all times is a simple way to prevent a number of hazards, and a dynamic pack limit will offer additional protection. The same strategy can be applied to the load device during discharging.

An interesting case is that of reading a zero, or nearly zero, cell voltage. Nearly all lithium-ion batteries are overdischarged at voltages this low, and therefore if the cell voltage is truly zero, most certainly an overdischarge condition exists. However, it is also possible that the same measurement is the result of a disconnected or broken interconnect lead. There are methods that are effective at distinguishing between these two events in certain cases, and in the case of a broken interconnect lead, it may be possible to retain some battery functionality.

If the battery management system allows for the measurement of the high-voltage stack at different levels of granularity, it is straightforward to determine if a zero-voltage reading is simply the result of a broken connection. If a measurement with enough accuracy is available of individual modules (substrings), groups of cells, or even the entire pack (although with large packs, the accuracy available usually does not permit the detection of a single cell at 0V). The sum of individual cells can be continually checked against the overall substring voltage to determine if errors occur. A broken connection will simultaneously introduce a mismatch between individual cells and the overall module voltage as well as creating one or more 0-V measurements. Because the broken-wire case is less severe (and in many cases, more likely) than a short-circuited or fully discharged cell, special detection for this failure type can prevent a battery shutdown and loss of functionality.

In the event of a lost connection with redundant measuring devices, if the primary and secondary measurement circuits are located in the same electronic module, connected by the same damaged connection, both measurements will be lost. In this case, it is impossible to ensure through the monitoring circuit that an overcharge or overdischarge condition does not exist in the disconnected cells. However, a few straightforward methods can be used to improve system performance while maintaining a high degree of safety. For applications such as electric vehicles in which battery charging through regenerative braking can be disabled, operating the battery in a discharge-only mode can prevent overcharging (as no charging is permitted). Overdischarging remains a risk if the unmeasured cells are appreciably lower in SOC than others and a significant margin of safety is not maintained between ending the discharge and entering overdischarge. If the aggregate state of charge error and total pack imbalance are bounded to a total of 5% SOC, then even if the most-discharged cell's voltage is unknown, if the aggregate SOC does not fall below 5%, then even unknown cells will not be at risk of self-discharge. Adding an appropriate margin of safety and reducing discharge limits will further increase safety. If it is not possible to disable charging while maintaining the battery operational, a complementary strategy could be used at high SOC, with higher risks as there is a higher hazard level associated with overcharge than overdischarge. This strategy is an effective method of maintaining reduced or limp-home functionality, but the need for safety is obviously paramount.

It is important to note that a cell that is shorted will promptly reach 0V. This is obviously a severe hazard and any use of these techniques must be accompanied by appropriate analysis to ensure that a short of one or multiple cells is not ignored by the battery management system. The maximum expected $dV/dt$ in the event of a short circuit should be known. Voltage changes exceeding this rate are likely to be due to poor connections to the monitoring circuit.

For most cells (the most negative and positive for each measurement circuit excepted), the loss of an interconnect lead will produce two adjacent zero or nearly zero voltage readings. If this condition appears spontaneously in conjunction with a voltage mismatch, a possible lost connection is the most logical diagnosis. Figure 17.2 shows lost connection detection.

### 17.2.2 Over-Temperature

In the same way as voltage rates of change are limited, the heat capacity and expected heat generation will limit the rate at which temperatures in a large-format battery system can change. However, in the event of a thermal event, the amount of available energy will be much higher leading to rapid temperature rises.

Rapid decreases in measured temperature that exceed the physical limitations of the cooling system (if there is one installed) can safely be disregarded as measurement errors.

Because over-temperature conditions usually lead to shutdown of the battery system, avoiding nuisance trips of over-temperature faults is critical. Therefore, a plausibility check on high-temperature events may be needed to improve system reliability. This must be weighed against the potential hazard of delaying battery shutdown in the event of an over-temperature event.

### 17.2.3 Overcurrent

The detection of excessive current exceeding published limits is discussed in Section 13.11. Very high currents should be prevented by a passive overcurrent protection device (generally a fuse or a circuit breaker) that requires no input from the battery management system to operate. The detection of a blown fuse is usually an important feature and can be accomplished by measuring high voltage on both sides of the fuse. Fuses react generally very quickly (especially if

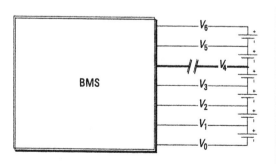

| Measurement | Actual | Measured |
|---|---|---|
| $V_{01}$ | 3.558 V | 3.558 V |
| $V_{12}$ | 3.601 V | 3.601 V |
| $V_{23}$ | 3.603 V | 3.603 V |
| $V_{34}$ | 3.604 V | 0.002 V |
| $V_{45}$ | 3.608 V | 0.004 V |
| $V_{56}$ | 3.602 V | 3.602 V |
| $V_{06}$ | 21.576 V | 21.576 V |

**Figure 17.2** Lost connection detection.

they are of the semiconductor type designed to protect power electronics), but all fuses clear more slowly with lower current, and larger fuses therefore allow more energy and higher current pulses before opening. Oversizing the fuse can result in other battery system components being exposed to potential overcurrent. In this case the system may rely upon the battery management system to prevent moderate overcurrents of sufficient duration to create a hazard but not long enough to clear the fuse.

The current sensor must have adequate measurement range to measure the applicable range of overcurrents. The increased measurement range will reduce the accuracy and precision of the current measurement during normal operation. The measurement system must also consider the possibility that sensor saturation occurs due to a sensor malfunction. Excessive discharge currents will be accompanied with significant voltage depression for all cells that can be used as a plausibility check for the high current.

The battery management system can command contactors to open to interrupt excessive currents. If contactors are to be opened under high current, they must be rated to interrupt the currents expected.

### 17.2.4 Battery Imbalance/Excessive Self-Discharge

If the battery cells begin to experience an excessive rate of self-discharge, the balance of the battery will become negatively affected if the differences in self-discharge cannot be compensated by the cell balancing system. Failures of the battery management system balancing circuitry are discussed in Section 14.5. In many cases excessive self-discharge will affect only a small number of cells due to a defect.

If the SOC and capacity for all individual cells are being computed, then determining imbalance and self-discharge is a relatively straightforward process. In the more common scenario in which the SOC and capacity information is calculated on an aggregate basis, other methods may be used to determine imbalance.

At a given aggregate SOC, the SOC for each individual cell should be expected to fall within a narrow range of $SOC_a - \varepsilon$ to $SOC_a + \varepsilon$. The slope of the SOC-OCV relationship is known throughout the SOC range, which can be used to determine a maximum expected voltage difference between cells. This approach can be used with care during battery operation but becomes more robust if the voltage measurements can be made while the battery current is zero (to avoid synchronization issues) and has been so for some time (to avoid differences in internal cell resistances and dynamics to influence the result).

A relatively simple check for self-discharge can be made even if the individual SOC of all cells is not accurately known simply using the relative rates of voltage decrease. If cells are close in SOC, then equal losses of charge should result in approximately equal decreases in voltage. If the battery system experiences long periods of inactivity where these can be measured (for example, in an electric vehicle that is turned off for many hours per day, or a backup power system that normally does not provide any current if the primary source of electricity is operational), the rate of voltage decrease can be simply calculated even with a single sample per day. Because all cells have seen the same current excitation, the internal dynamics should relax at the same rate. If a particular cell exhibits an excessively high rate of self-discharge, the voltage decrease should be evident after a few days of sampling if voltage measurements are accurate—appropriate lowpass filtering is needed, but this method avoids the complexity of calculating all individual cell SOCs and requires little sampling.

### 17.2.5 Internal Short Circuit Detection

This is a challenging area of active research. Many methods focus on the detection of a change in the shape and/or duration of the charging taper or in detecting noisy voltage signals which occur as the cell approaches 100% SOC at low current.[1]

### 17.2.6 Detection of Lithium Plating

This is a relatively new concept that has been discussed in academic literature [2]. If metallic lithium is deposited in a lithium-ion battery due to abuse, due to the lower anode voltage created by the lithium anode, the shape of the OCV/SOC relationship will change on discharge with a higher-voltage plateau appearing at the start of a discharge cycle. Low discharge currents, which could be possibly created by the balancing circuit, will show a higher cell voltage than cells without lithium plating. Significant developments are needed to ensure this concept is robust, but detecting this condition is desirable, and likely this will become a feature in many large-format battery management systems in the future.

### 17.2.7 Venting Detection

Recently, sensors have become available designed to detect volatile organic compounds in electrolyte. The presence of volatile organic compounds in a battery system is often associated with cell venting and may provide an early warning of this hazardous event.

These sensors can detect cell venting regardless of the cause (over-temperature, manufacturing defect, mechanical intrusion, overcharge) and pro-

---

1. There is little public evidence of commerical implementation of this technology, but refer to [1] for possible techniques to detect.

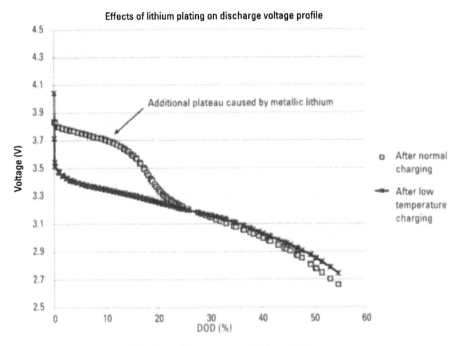

**Figure 17.3** Voltage profile affected by presence of plated lithium.

vide improved safety in the event of cell venting. These technologies are still relatively new and should not be relied upon for primary protection.

### 17.2.8 Excessive Capacity Loss

An important distinction should exist between normal loss of capacity due to again phenomena and excessive, rapid loss of capacity due to a battery defect. An effective capacity detection scheme can be coupled with an expected worst-case capacity trajectory to determine if the battery is declining more quickly than usual. Often capacity loss will be due to unmanageable self-discharge in one or more cells. Significant and unexpected capacity loss may indicate a serious internal defect due to internal shorts, cell damage, or lithium plating.

## 17.3 Reaction Strategies

When a fault occurs, the battery management system should bring the battery to a safe state as quickly as possible, while maintaining an appropriate level of functionality. The definition of all of these reactions is specific to the nature of the battery system as well as the specific type of error condition being addressed.

Fault strategies should always take a secondary role in protection if primary protection methods, such as limit algorithms fail to protect the battery from hazardous conditions. The various systems of the battery management system should be designed such that if the load respects the limits published by the battery and the battery SOH is acceptable, fault or error conditions should not occur. Faults should be associated with a system malfunction, unexpected battery response, or degradation or unexpected load condition.

A common strategy for serious faults that do not require immediate disconnection of the battery, but need a timely response, is to allow limited operation for a specified period of time, but to prevent restarts or reconnection of the battery system after this period has passed. This can allow adequate energy for the load device to execute an orderly shutdown.

In the event of impending overcharge or overdischarge in which the load has failed to respond to requests to reduce battery current to zero, the battery can simply be disconnected from the load to prevent further changes in battery SOC. Although this will prevent further charging or discharging, damage to battery system components may result.

Stationary energy storage systems often incorporate fire suppression systems that could be activated by the battery management system. The operation of any last-resort safety systems such as this should have appropriate measures such as requiring at a minimum, no single-point failures in hardware or software will lead to inadvertent actuation.

## References

[1] Mikolakezia, C., et al., "Detecting Lithium-Ion Cell Internal Faults in Real-Time," *Power Electronics Technology,* March, 2010.

[2] Zimmerman, A. H., and M. V. Quinzio, "Lithium Plating in Lithium-Ion Cells," Battery Workshop, Nov. 2010.

# 18

# Hardware Implementation

## 18.1 Packaging and Product Development

The final battery management system components will often incorporate enclosures to provide protection. Enclosures improve the ability of electronics to resist electromagnetic disturbances, heat, humidity, pollution, and other damaging environments but also add cost, mass, size, and complexity. A variety of techniques are used for battery management system devices depending upon the application. The following considerations can help select an appropriate package for the chosen design.

Positive locking connectors with many terminals (common for voltage sense leads for large-format battery packs, see Figure 18.1) can require significant mating forces when installing them. These may cause distortion of circuit boards and may lead to temporary breakdown of clearance distances. In many electronic systems, these effects are negligible due to their temporary nature, but since the battery stack is always live, this could result in an arcing hazard when the connectors are mated. Ensure that connectors are well supported and that mating loads are considered.

The use of liquid encapsulants ("potting" compounds) can improve the environmental performance of electronic devices. Performance of these materials is highly dependent upon application. For critical applications such as battery management systems, consider automated application and inspection if encapsulants are relied upon.

Various polymers are susceptible to considerable water absorption that may reduce their dielectric performance.

**Figure 18.1** High-quality connector for automotive applications. (Courtesy Molex.)

The battery management system devices may often require appropriate warning labels to warn users of hazardous voltages.

## 18.2 Battery Management System IC Selection

The last few years have seen the introduction of at least a dozen specialized integrated circuits that are intended for use as building blocks of a large-format, lithium-ion battery management system. Although it would seem that these devices are all comparable and are the ideal solution for implementing the basics of cell voltage measurement, they are not all created equal and their use requires as much care as an implementation built from discrete components.

The selection of a monitoring IC is a critical decision in the battery management system implementation. At the time of this writing, a large number of semiconductor manufacturers were offering first-generation devices with a few providers recently releasing second- and third-generation products. Because there is little standardization in these devices and proprietary techniques are used for communication between devices and from device to host processor, if a particular device is shown to not meet the requirements during the process of development and validation, switching to another manufacturer's IC will not be a simple matter of swapping one part for another late in the design process. These components do not offer the same measurement topology, level of accuracy, sampling rate, or safety features, nor do they all measure the same number of cells or communicate with other devices the same way. Choosing a chipset must be done with great care to avoid last-minute design changes.

Some of the cell measurement ICs have been augmented with the availability of secondary measuring devices for performing redundant measurements to provide an architecture with a high level of functional safety. Many of the features of the devices are comparable but some important differences exist.

These chips almost invariably require a high-voltage semiconductor process. As these processes are specialized and relatively new, suppliers should be evaluated on their level of experience and familiarity with the process they have selected.

The measurement topology used in cell measurement ICs varies considerably with a number of parameters to consider. The most fundamental attribute is whether an individual ADC is used with each cell or whether a single ADC is multiplexed across multiple cells.

Most of these devices use a level-shifted or current-source communications bus where the ICs can be daisy-chained together and connected to the series stack of cells. Each IC communicates with the next higher voltage IC (connected to cells at a higher voltage) and lower voltage IC over this bus, and the lowest IC in the chain is connected to a host processor. An example of how these devices could be used is shown in Figure 18.2.

The failure modes associated with monitoring ICs should be considered. A number of manufacturers make a failure modes and effects analysis (FMEA) available to customers upon request. Additionally, reliability data (specifically, FIT ratings for IEC 61508/ISO 26262 latent fault analysis) are necessary to meet the requirements of many safety standards, and manufacturers should be prepared to discuss the level of testing and calibration that accompanies the devices. Many applications and customers demand 100% testing for critical parameters.

A survey of some of the available devices is found next.

### Linear Technologies Corporation (LTC)

Linear Technologies offers the LTC680X family of devices, including the LTC6801 and LTC6802, LTC6803, and LTC6804. The LTC6802 was one of the first devices released for this application and was quickly superseded by the second-generation LTC6803, offering a number of additional features. The LTC6804 is a recent development that offers a proprietary isolated SPI bus. The LTC6801 is a complementary secondary measurement device that performs single threshold overvoltage and undervoltage detection. All devices are capable of measuring 12 cells. These devices offer high accuracy but a significant cost drawback is the requirement of the capacitor recommended between each cell and the negative terminal of the lowest potential cell connected, requiring capacitors with 60V or 100V ratings. The LTC6804 is a third-generation component offering very high accuracy (1.2-mV maximum measurement error) and a fast synchronization between all measurements with a programmable noise

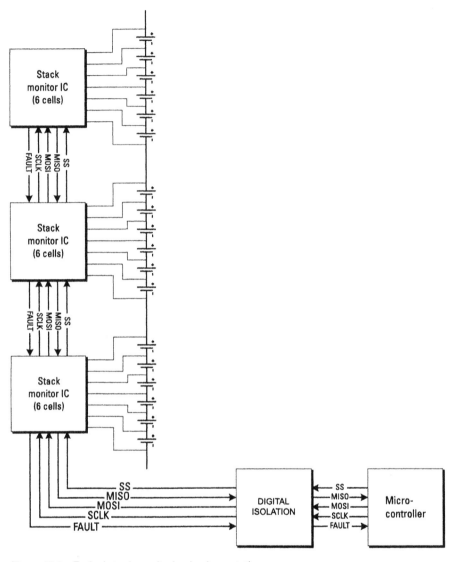

**Figure 18.2** Typical stack monitoring implementation.

filter. Sampling rates can be reduced to improve noise performance. The devices support passive cell balancing and allow the use of the cell stack or an isolated voltage source as a power supply.

### Analog Devices

Analog Devices provides the AD7280A lithium-ion battery measurement IC as well as the complementary AD8280 hardware-only monitor device. Each AD7280A device measures six cells using a multiplexed topology into a single

12-bit SAR-type ADC. The device offers 1-$\mu$s per channel conversion time and claims ±1.6-mV accuracy for cell measurement as typical, but the worst-case ratings are between 9 and 14.5 mV depending on operating temperature range. The communications architecture is a level-shifted SPI bus. Analog Devices explicitly discourages the use of this level-shifted SPI between PCBs or modules in a battery system. Each individual channel is capable of measuring between 1V and 5V, which is a higher minimum voltage than other devices available. Each AD7280A offers six general-purpose ADC channels that can be used for temperature measurement which is a higher number of temperature channels per voltage channel than most devices offer.

The AD7280A has a cell-balancing output for each of the six cells which can be used to drive the gate of an external FET for cell balancing purposes.

The AD7280A and AD8280 are powered exclusively from the high-voltage battery stack. Various modes exist for reducing power consumption. The AD7280A devices typically consume between 5.1 and 6.9 mA from the battery stack during operation and between 2.5 and 3.8 mA during software power-down. Maximum ratings are 30% higher. No specification is given for the variation between devices. In full power-down mode, these devices consume no more than 5 $\mu$A. For a typical 100-Ah lithium-ion battery pack with a self-discharge of 2.5% per month, the AD7280 alone will double or triple the apparent rate of self-discharge if the devices operate continuously. The AD8280 adds an additional 2.0 mA worst case to the stack consumption in active mode and 1.0-$\mu$A in shutdown mode.

The 48-pin LQFP package has a 0.5-mm lead pitch with as little as 0.23 mm between pins. While the individual cell measurement inputs are spaced appropriately to minimize potential difference between adjacent pins, the device requires $V_{DD}$ and $V_{SS}$ inputs that are curiously located immediately next to one another. These two pins could be as much as 30V apart and the clearance distance between the two pins is minimal. Analog Devices recommends the use of a coated circuit board, but care should be taken to ensure that the clearance and creepage requirements of the application are met.

### Texas Instruments (TI)

TI's bq76PL536A IC offers high-accuracy measurement of between three and six cells per IC, with a high-accuracy 14-bit ADC, which claims 1-mV accuracy for cell measurement (worst case of 2.5 over a temperature range of −10°C to 50°C and 5 mV over the entire temperature range of the device). The bq76PL536A is a second-generation device offering improvements over the bq76PL536. This is significantly better than other available devices and is a good candidate for lithium iron phosphate systems requiring accurate cell measurement. The integrated successive-approximation register (SAR) ADC has an internal bandgap voltage reference and performs the A/D conversion in 6 $\mu$s per channel. The device includes two temperature measurement channels and also measures the six-cell substring voltage as well as a single general-purpose analog input allowing for robust broken-wire detection. The bq76PL536A incorporates internal secondary protection for overvoltage and undervoltage as well as overtemperature. A pair of hardware outputs (for alert and fault levels) is used to report the presence of these faults, eliminating the need for a host microcontroller to be involved to respond to these events. The thresholds and delay times for the undervoltage, overvoltage, and over-temperature conditions are stored in on-chip ECC flash that is robust against but not immune to data corruption. Each cell offers a single control line for passive cell balancing through the use of an external resistor and FET. Each device is factory-trimmed to compensate for measurement errors on all inputs. The TI IC uses a hardware input that can be used to synchronize measurements if using the firmware commands does not provide the desired level of synchronicity. The conversion trigger signal is carried in the level-shifting interface to the next higher device. The device requires only a single 1-k$\Omega$ resistor and 1-$\mu$F capacitor on each cell voltage input and the "ground capacitor" configuration requiring high-voltage capacitors is not specified. These devices should be high precision and high quality to maintain accuracy, but the external components required are less costly than with other manufacturer's ICs. The communications interface is once again level-shifted SPI, using a digital signal to trigger the conversion.

The TI device has a 36-V peak stack voltage rating allowing continuous voltages of 5V per cell with tolerance to peaks of 6V per cell.

The device is capable of performing each cell measurement within 6 $\mu$s. The relatively small number of channels per device means all cells are measured in worst case of 42 $\mu$s (6 × 6 + 6 $\mu$s for the start of the conversion).

A potential shortcoming is the device's secondary protection features are disabled if the device goes into thermal shutdown. The device also uses one-time-programmable (OTP) EPROM to store the critical calibration parameters for the secondary protection device, meaning they cannot be reconfigured through software reflash (either accidentally or intentionally) once the device is deployed in the field. OTP programming also requires either additional on-board circuitry or provisions for programming the devices before they are attached to the board with possible damage to the components. The device does not allow for an external voltage reference although the primary and secondary measuring circuits use separate references. The internal reference voltage cannot be used to drive any external components. The hardware over-temperature protection relies upon a particular type of external 10-k$\Omega$ thermistor that although common may not be suitable for applications requiring temperature cut-through protection. If the over-temperature function is not used, then any type of thermistor can be selected.

Sleep current is 12–22 $\mu$A, which is higher than competitive devices. Activation of the secondary protection circuit requires 46–60 $\mu$A, and 10–15 mA are drawn during conversion. Additional current is drawn when one or more signals in the level-shifted interface are asserted.

The device uses a TQFP-64 package for each six cells requiring more real estate than the AD device. The lead pitch is still 0.5 mm, but high potentials between adjacent pins are minimized.

The single device implementation of the primary and secondary monitoring reduces cost and complexity at the increased risk of a common failure mode that damages both circuits.

### Atmel

Atmel offers currently the ATA 8670 IC. Each IC is capable of monitoring six cells but only uses a 30-V CMOS process, compared with other devices using higher-voltage processes. The maximum string rating is 30V, allowing for only 5V per cell. Each device also allows for two temperature measurement channels, and only allows for one temperature to be measured during each measurement cycle. Due to the individual ADCs per measurement channel the device is capable of simultaneous voltage measurements, which is a unique offering uncommon in the marketplace.

The device contains the capability for dissipative cell balancing using an external transistor and discharge resistor.

Unlike other devices, the Atmel product uses an individual ADC per cell monitoring channel and does not rely on a level-shifting analog multiplexer. Each ADC is 12 bits. The maximum error is ±20 mV.

### Maxim

Maxim offers the MAX 11068, MAX17830, MAX11080, and MAX11081 devices. The MAX 17830 is the second-generation recommended replacement for the MAX11068, which is no longer recommended for new designs, but details on the new component are still not publicly available. The MAX11068 and the replacement MAX17830 are each capable of monitoring up to 12 cells with a cell voltage between 0V and 5V per cell. The devices are built using an 80-V semiconductor process.

The interface between devices uses a level-shifted I2C protocol in comparison to most other devices using SPI.

In standby mode, the 11068 devices consume 75 $\mu$A and only 1 $\mu$A in shutdown mode.

The MAX11068 offers integrated cell balancing switches capable of switching up to 200 mA, unlike many other devices, including its suggested replacement MAX17830.

The MAX11080 and MAX11081 are used for secondary protection and provide over and under-voltage detection. The devices offer fixed hysteresis of 300 mV (MAX11080) or 37.5 mV (MAX11081) with a fault delay time programmable using an external capacitor. Typical accuracy for overvoltage detection is ±5 mV (max 25 mV) and 20 mV typical, 100 mV maximum for undervoltage accuracy, the latter figure of which is somewhat limiting.

A simple RC filter is required for each input to the MAX11080/11081 using a 0.1-$\mu$F/80-V capacitor (bringing the cost disadvantages of high-voltage capacitors) and a 10-k$\Omega$ resistor. The MAX11080 and MAX11081 have a number of pins that are rated to handle a 2-kV HBM ESD event.

For high-volume OEM applications, lithium-ion battery stack monitoring is an active area of development for many semiconductor manufacturers. A number of devices exist that were not, at the time of this writing, disclosed to the general public except to end users who could demonstrate significant interest due to intellectual property concerns. Extensive consultation with stack monitor IC vendors is encouraged to understand the latest products that may still be in the development cycle. Manufacturers who are offering first-generation products should be looked at with additional scrutiny to ensure that pitfalls encountered during the launch of other competitive devices are not repeated.

## 18.3 Component Selection

### 18.3.1 Microprocessor

The criteria for microprocessor selection for battery management system applications are similar to those for any safety-critical control electronics application. Many manufacturers offer a wide range of products which can be used to build an effective battery management system. Some important considerations are noted next.

Safety-critical systems should be robust against random corruption of microcontroller software. In standard SRAM, the probability of a single memory bit being changed due to random interaction of the microcontroller with the environment [known as a single-bit upset (SBU)] is approximately 2,000 failures per $10^9$ operating hours. The use of an MCU with ECC (error-correcting) memory will reduce the effects of these types of events.

Many functional safety-rated designs in the past have relied on the use of multiple MCUs to achieve the required levels of safety for functions associated with high hazard ratings. At the time of this writing, a number of vendors were offering all-in-one designs that, according to the vendors, were rated for use in these applications.

A number of microprocessors incorporate advanced features for functional safety, including multiple cores, lock-step operation, and error-correcting internal memory and offer safety-rated MCUs for safety critical applications such as battery management systems.

### 18.3.2 Other Components

Connectors are a particularly sensitive area for high voltage failures. Many connectors are limited in the voltage ratings allowable and do not provide for separated cavities in both halves of the connector system for individual terminals. Connectors without this feature are extremely prone to pin-to-pin shorting if the connectors are not inserted with great care. D-subminiature computer connectors are a particularly good example of this and should not be used for any high-voltage application due to the possibility of short-circuit and associated damage.

To meet creepage and clearance distances, it may be logical to place high-voltage inputs on a separate connector with larger pin-to-pin spacings and use a more conventional connector for low-voltage control and power signals. Figure 18.3 shows a connector pin layout that reduces the potential difference between pairs of adjacent pins.

In the situation in which many connectors are used in a minimally distributed design to bring cell voltage sense connections to a battery management system module, consider the possibility of connecting these connectors incorrectly and applying unexpectedly large or negative voltages across cell sense connections. While the use of a larger number of connectors may reduce risks associated with the maximum voltage present in any one connector or bundle, it does increase the risk than an incorrect connection occurs. Connector keying is available with many series of connectors up to a limited number of configurations and should be used for all high-voltage connections.

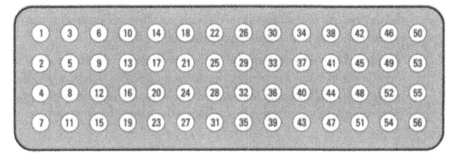

**Figure 18.3** Effective connector layout for reducing potential difference on adjacent pins.

The voltage value of small surface mount resistors is rarely considered (as compared to capacitors, which are normally selected by voltage rating) and may be very low (less than 50V) in relation to the voltages encountered inside a battery management system. The voltage rating often has little to do with the resistors power rating and resistance value, meaning that achieving the rated power may actually be impossible without exceeding the voltage rating.

## 18.4 Circuit Design

An important thing to consider is that the connections to the high-voltage battery stack are always live. These connections will be exposed to continuous high voltages for months or years of operation with no or very few instances of power-down. Large-format systems are capable of delivering sufficient current to cause fires in electronic modules if short circuits develop. Many electronic circuits are not subjected to this type of operational requirement. This is also true of many control voltage inputs, but the hazards associated with the battery stack are much higher due to the voltages involved. Consider the effects of any component failure that may lead to a short circuit or a "soft" short across the high voltage battery stack, between individual cells or across groups of cells. Filter capacitors or TVS diodes can fail in a short-circuit mode and lead to a fire.

Common techniques to avoid this hazard include:

- Uprating of certain components to reduce the risk of a short circuit by providing a higher margin between the working voltage and the maximum rating.
- The use of multiple capacitors in series to ensure that a short-circuit failure of a single device does not lead to a short circuit across a high potential. As the failures that lead to shorted capacitors are often mechanical in nature (due to mechanical shock, stress, or vibration), a good practice is to place the capacitors at right angles to one another to ensure that the devices are not subjected to the same mechanical loads that would presumably lead to the same type of failure.
- The use of a crowbar circuit can be used to prevent a disastrous event at the expense of usually rendering the device inoperable. A fuse is installed between the high-voltage source and the component with a potential short-circuit failure mode, which clear in the event of a short circuit.

It is often important that a battery management system is capable of being connected to battery cells with the control power disconnected (for systems using separate power supplies). For a number of semiconductor devices, this can present the risk of CMOS latch-up. If a voltage is applied to the inputs or outputs of a CMOS device before the power supply is present, an internal short can occur that usually destroys the device. Because the cells always present a voltage on connection, there is a requirement to use a topology that prevents CMOS latch-up. A thorough test plan for these conditions is also warranted. ICs connected directly to cells should be constructed using either a trench technology or using silicon-on-insulator (SOI) technology that is resistant to latch-up, or they must be appropriately protected using external devices that prevent CMOS devices from latch-up.

There are risks associated with the use of PCB traces as fusible links. Consistent clearing current for a fuse depends on precise control of the conductor's cross-section and thermal resistance to the surroundings that are not assured by the PCB manufacturing process. Additionally, the fusing of a trace may ignite the PCB substrate, or spatter copper over other components leading to other potential problems. Inexpensive surface-mount fuses can be used for this purpose instead.

Additionally, an important corollary of the above-mentioned issue is problems arising during battery management system installation if all battery cells are not connected to the measuring circuit simultaneously. In most cases using standard wiring techniques and connectors, the connection order is impossible to guarantee. A number of ICs have developed internal protection circuitry to address these two issues, but this remains a failure mode that should be thoroughly verified by analysis and testing. Combinations of different types of ICs for primary and secondary monitoring could reintroduce this failure mode at the module level. The use of external clamping diodes can be used to ensure that device power rail voltages are never exceeded by the voltage present at the inputs.

In many cases battery management system components are powered from the high-voltage cells themselves. In this case, it is important to consider the voltage range that the devices will experience throughout the full SOC range in which the cells will operate, both under normal and abnormal operations. If a cell falls to a very low voltage due to self-discharge or misuse or is overcharged, connected components could be operated outside of their maximum allowable ratings. While it may be acceptable that accurate voltage readings are no longer possible during severe overcharge or overdischarge (including the possibility of cell reversal), it should be ensured that circuit components do not overheat or produce erroneous results when exposed to extreme cell voltages.

For circuits that could lead to high-voltage short circuits or potential isolation or ground faults, a thorough component-level failure mode analysis should be performed

## 18.5 Layout

As discussed earlier, layout activities should ensure that the required creepage and clearance distances are maintained on the PCB assembly.

Maintain the high-voltage circuits in a segregated area on the PCB and delineate all isolation barriers cleanly through all layers, preventing the possibility of a layer-to-layer fault. Consider that placing high-voltage components near the edge of the PCB increases the possibility for breakdown between components and enclosures or mounting features.

## 18.6 EMC

The presence of switching power electronics, high currents and power levels, and sensitive analog measurements poses a significant challenge to achieving electromagnetic compatibility for battery management system products.

An area of EMC design and testing that is often overlooked in battery management system development is both the susceptibility and emissions of the device to dc and low-frequency magnetic fields.

The dc magnetic fields are produced by dc flowing in busbars, cables, and other interconnects, as well as the batteries themselves. The battery current in most large-format systems can be very high (hundreds of amperes or more), creating strong fields. Many electronic systems are not exposed to this intensity of dc magnetic fields. Additionally, battery management systems may contain or make use of Hall effect or magnetostrictive current sensors that can be influenced by static fields.

The high-current path of the battery pack is often connected to switching power electronics that may cause conducted emissions that would occur at the switching frequency of the electronics or above due to harmonics. This could cause strong ac fields with a dc offset in the 12–200-kHz range (lower-frequency range than most EMC testing), which could potentially interfere with management electronics.

The numerous voltage inputs for the battery cell stack connected to the battery management system are difficult to simulate with any type of power supply, during EMC testing. Consider using smaller-format battery cells or the actual battery system.

EMC testing must consider the high-precision, high-accuracy nature of battery management system components. Many EMC tests look only for gross errors and failures. Erroneous voltage measurements may cause a battery management system to misreport the battery capability and lead to a failure. While under severe disturbances it may be acceptable to experience errors in voltage and current measurement, the battery management system software should be capable of performing plausibility analysis on rapid fluctuations in voltage and current and take appropriate actions. If the battery is operating near nominal parameters (no extremes of SOC, temperature, voltage, or current), it is acceptable to ride through a short disturbance. For longer disturbances or for battery systems operating closer to their limits, system shutdown may be the only way to ensure safe operation.

The routing of the high-current path, sensor position, and orientation are important for accurate reading under strong field conditions.

Voltage measurement inputs that, by design, should be high impedance are sensitive to current-mode noise due to current injection. Even small currents can cause large voltage errors.

An important area in which few EMC/EMI standards exist is the interaction between the low-voltage and high-voltage systems in cases in which the two are isolated. The safety issues associated with Y capacitance have already been discussed. There is also a significant system performance issue associated with precision measurements in isolated systems. In an ideal situation, the two systems are completely isolated from one another, but there are many deviations.

- Isolators that are nearly always incorporated in the battery management system introduce a very high (but not infinite) resistance between the two sides of the isolator, and a low (but not zero) current can flow across the isolator during system operation.
- Capacitive linkage also exists between the two sides in the form of both intentionally installed Y-capacitors as well as parasitic capacitances.
- Ground fault measurements, which may be made on a regular basis, may insert a measurement resistance between the high-voltage stack and the earth ground.

These three conditions can combine together to create performance problems for various types of measurements. When a ground fault measurement is made, the presence of Y-capacitances can interfere with the ground fault measurement during the time that the RC circuit formed by the Y-capacitances and the measurement circuit is charging (see Figure 18.4).

## 18.7 Power Supply Architectures

If a sleep mode is implemented in which the microprocessor is powered down to reduce power consumption to an extremely low level, hardware will be required to implement a transition out of the sleep mode. The most common condition requiring this is a command to connect the battery to the load, which usually is provided via a digital input signal or a network command.

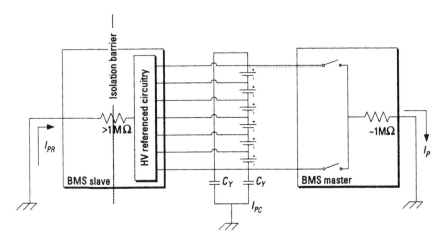

Current injected ($I_P$) when isolation measurements are made and establish reference potential between HV and LV systems. Injected current across isolation barrier can create voltage measurement errors.

**Figure 18.4** RC circuit formed between Y-capacitance and measurement circuit.

Sources of battery management system wakeup may include:

- Certain network messages may trigger a hardware wakeup while only specific commands may lead to the battery management system becoming an active node and transmitting.
- Digital signal states or transitions from the load device requesting battery actions.
- Digital input pins may be used to wake the battery management system in special cases, such as bench testing when the battery management system is not connected to a battery system.
- Periodic verifications of the battery state triggered by an internal real-time clock or timer circuit.
- Connection to an external charger.

## 18.8 Manufacturing

The safety-critical nature of battery management system components requires that the due care taken in their design is not undone by defects introduced in manufacturing.

The highest standards of workmanship are often warranted for battery management system manufacturing. IPC Standard 610, Class 3, is an example of an appropriate specification for acceptable workmanship for printed circuit board assemblies destined for battery management system application.

Particular concern needs to be paid to accuracy of trace spacing and trace widths that could compromise creepage and clearance distances due to layer misalignment. Panels should be fully electrically tested to ensure there are no shorts or opens before the panels are populated with components. Reworked or "X-out" panels should not be accepted. A full process FMEA (PFMEA) should be conducted by a coordinated team with detailed knowledge of the impact of various types of process failure modes on the ultimate overall battery system in the proposed application.

Tin whiskering is a problem that has come to light in the recent switch to lead-free solder processes for the manufacture of many electronic components. The growth of tin whiskers on PCB surfaces can cause shorting of components at high relative voltages and lead to thermal events within the battery management system or the battery system. Manufacturers using lead-free processes should be evaluated for their experiences in avoiding tin whiskering. Tin whisker growth is promoted by exposure to high temperatures and high humidity, and an appropriately designed test plan can be used to determine to what degree this hazard has been eliminated.

The level of care does not end with the completion of the printed circuit board. Enclosures are a potential hazard. Dented or misshapen enclosures can reduce creepage and clearance distances or even contact hazardous voltages. Loose fasteners or broken pieces inside of a module are a serious short-circuit hazard. Sealing may be compromised due to missing or poorly installed seals, enclosure damage, or process variation, especially with liquid-applied sealing products. Use of thread-cutting or thread-forming screws to improve manufacturability must be carefully reviewed to ensure that metallic debris is not introduced as these fasteners are installed.

# 19

# Software Implementation

Software in a modern battery management system is required to perform complicated estimations of internal battery parameters according to advanced models and algorithms, and also to react quickly and decisively to safety hazards which may arise within the battery system. The expectation that the software should make these decisions in the proper way in a timely manner must be met if battery systems are to perform well.

Software can be generalized into the transformation of a given set of inputs into a desired set of outputs. In a real-time system, this processing must not only occur correctly each time, but in a maximum allowable time. Reacting too late, in the case of certain events, may be as ineffective as not reacting at all.

Examples of time-critical events that modern battery management systems must correctly process include:

- Automated battery disconnect during the crash of a battery powered vehicle;
- Disconnection of the battery in the case of an interlock being opened;
- Reducing power limits as the battery approaches the edge of the safe operating area.

Determinism is a key component of real-time, safety-critical systems. Fault checking should occur on a regular basis. If a problem is detected, it should be guaranteed by analysis of the design that the reaction will occur in a fixed maximum time, regardless of the conditions experienced by the battery management system. A number of conditions can degrade the responsiveness of the software that can be overlooked if the developer is not familiar with real-time systems.

The use of interrupts can interfere with the deterministic execution of software code. Although they are indispensable in modern system implementations, they must be used with caution to prevent problems. Use hardware-triggered interrupts with extreme caution. Hardware interrupts allow circuits outside of the microprocessor to change the control flow timing in a way from which the software may not be able to recover. Appropriate hardware filtering or glitch removal should always be used if this type of hardware interrupt cannot be avoided. Nested interrupts with priority levels are another possible source of unstable execution timing.

As a rule, keep interrupt service routines short, few in number and with little to no decision making. In general, an interrupt service routine should be executed to respond to the arrival of a piece of information; this information should be retrieved to make room for the next one and placed somewhere to be processed later as part of the normal deterministic program execution.

Overloaded or flooded communications buses are a common cause of software performance errors. Interrupt-driven message retrieval routines should contain "safety valves" to prevent excessively high rates of execution. Although the loss of messages could prevent proper system operation, it is usually preferable to losing time-critical performance or possibly preventing the software from reacting in real time. Testing at the worst expected normal and overload conditions is beneficial and can demonstrate that the system can recover in a safe way.

## 19.1 Safety-Critical Software

The safety level required will influence the selection of the programming language (or in some cases, the programming language subset) used in the development of the battery management system software.

The C language, despite being the de facto standard for many industries and applications, has a number of shortcomings for safety critical applications. Access to low-level memory, dynamic allocation and lack of strong typing can lead to latent software defects that do not show up in routine testing which may have unpredictable results.

The Motor Industry Software Research Association (MISRA) recommends a subset of the C language for automotive and other safety critical applications. MISRA C prevents a number of constructs and operations determined to be dangerous. These operations may have unintended side effects that are unlikely to be caught, may reduce the determinism of the software, or may cause issues such as memory leaks that lead to unpredictable software failures.

As an example, the use of dynamic memory allocation, common in the development of software for desktop PC applications, should be avoided in safety-critical embedded systems. Although dynamic allocation is useful for applications in which large arrays of data of variable size must be managed (such as spreadsheets or databases), data of this type is not often encountered in embedded controllers. Memory management errors such as memory leaks, overwriting of critical information, or running out of memory can cause disastrous software faults and lead to application crashing or memory violations that cause corruption of data that is used to make safety-critical decisions. The use of advanced techniques such as dynamic allocation, heavy use of indirection (pointers), self-modifying code, or functions with multiple entry and exit points all have significant risks that should be avoided.

Battery management system software should be considered safety-critical. Safety depends upon timely reactions from the battery management system if hazardous battery conditions occur. It may be possible to decompose the software into sections that are safety-critical and those that are not. In this type of a system, software components responsible for fault detection and reaction, voltage and temperature measurement, current measurement, and isolation detection are likely to be critical for system safety. Wheather functions such as SOC and limits calculation are considered safety-critical will be application-dependent.

## 19.2 Design Goals

A common design goal of safety-related systems is that their design should be in accordance with the ALARP (as low as reasonably practicable) principle for risk. This implies that risks should be reduced to the lowest practical level. This does not mean that risks can be completely eliminated. Nor does it mean that the most risk-adverse technical strategy is chosen. It implies that the costs (financial and otherwise) of avoiding risks must be weighed against the severity of the hazards associated with the risk. Unless the costs are disproportionately large in comparison with the benefits of eliminating risks, the risks should be avoided.

It is important that battery management system designers can demonstrate that a level of due care has been taken and that the product design and implementation reflect a level of safety that is appropriate to the application.

## 19.3 Analysis of Safety-Critical Software

A number of techniques and tools exist for the analysis of software and should be performed on applications considered to be safety critical.

Peer review is the most basic form of analysis or inspection. All embedded control systems should encourage peer review of software, at least on sections of the code that are critical. A formal method of code review should be established by teams developing battery management system software.

Static analysis tools such as LISP or QA/C can inspect code for a number of types of potential risks. These tools can be used to ensure compliance with coding standards such as MISRA, to locate potentially dangerous constructs with unwanted side effects or that may execute in a way unintended by the coder. The use of static analysis tools is strongly suggested by standards such as ISO 26262 as well as the internal requirements of many OEMs in various markets.

## 19.4 Validation and Coverage

It is impossible to simulate a modern embedded system with a full array of test vectors that cover every possible situation that may be encountered, but battery management system validation should follow a rigorous process that ensures that validation cases exist for every requirement, especially those associated with safety-critical functions.

Obviously many of the test conditions that are required involve replicating situations that are unsafe to reproduce with live batteries and therefore an effective simulated battery pack environment is necessary to fully test the battery management system. A surrogate test system is required to replicate the voltages, temperature, and current signals, as well as any messaging networks and other inputs and outputs required to reproduce the environment in which the battery management system operates.

There are relatively few options for a standardized testing platform for battery management systems, and the equipment requires is often highly specialized. The ideal test environment contains the necessary hardware to:

- Reproduce battery cell voltages with a high level of accuracy (as a rule of thumb, test equipment should have four times the precision of the device under test). With most battery management systems offering accuracy in the 1–10-mV range, test equipment should be able to produce cell voltage signals with 0.25–2.5-mV total error). These cell voltage signals will also need to be able to be strung together in series to create a high-voltage stack as they would in a real battery pack, requiring an appropriate level of isolation between the outputs and the test system ground. In addition, because the battery management system will be performing cell balancing as well, these channels need to act as a source (for dissipative balancing) and possibly a sink (for charge-transfer balancing) of current. These current ratings will need to be matched to the battery management system capability for cell balancing currents. The voltages should be individually addressable with an update rate at least two times the measurement rate of the battery management system sample rate. The term SMU (source measurement unit) is often used

for a voltage source capable of sourcing and sinking current, and which provides for accurate voltage and current measurement of the output.
- Provide accurate simulation of temperature and current as well. Simulating thermistors can be done with a variable resistance device or with a voltage source; the latter is usually less expensive but may not allow reproduction of short-circuit and open-circuit conditions.
- Reproduce dynamic battery conditions accurately. The voltage, current, and temperature measurements need to be driven by a model that reproduces the behavior of the plant (battery system) that is being simulated. These quantities need to update synchronously in real time as they would in a real battery system if the capabilities to perform real-time impedance detection, SOC estimation, and SOH calculations are to be modeled. The test model must operate with equal, or ideally higher, fidelity than the model implemented in the battery management system.
- Simulate fault conditions such as ground/isolation faults, broken wire, contactors that fail to open or close, short and open circuits in sensor connections, violations of battery limits, off-scale high and low readings, implausible combinations of inputs, as well as battery failures such as severe capacity and state of charge imbalance, self-discharge, excessive impedance and rapid capacity loss.
- Perform high-speed detection of battery management system responses to critical fault conditions to verify that the response time is adequate.
- Simulate overload conditions such as increased frequency and number of communications messages, overcharge/undercharge, and absence of slave devices.

It is important to realize that the battery management system test fixture will also incorporate high voltages and must therefore incorporate adequate safety, guarding, emergency stop, and insulation/isolation provisions.

While there are few standardized vendors or tools specifically for battery management system testing, a number of devices have become available to respond to the demand for automated battery management system validation.

Pickering Instruments offers a series of PXI form factor battery simulator devices that can simulate battery cell voltages. The devices are suitable for battery stack voltages up to 750 VDC and support 300-mA discharge/100-mA charge per channel to support cell balancing. Each device simulates six independent channels that can be connected in series to form a partial simulation of battery cells.

## 19.5 Model Implementation

When a valid battery model is developed, it must be implemented into the embedded controller used on the battery management system to estimate the battery state in real time.

This can be performed in a number of ways, including autogeneration of software code from a model-based development environment such as MATLAB/Simulink.

The scope of how the model will operate must be chosen. Ideally, the model can run on each individual cell that the battery management system measures to generate the SOC, the SOH, and power limits for each cell. These individual cell results can then be cascaded to create pack-level results.

The pack-level SOC will depend on the individual cell capacities, SOC, and system balancing capability. Assume that cell $n$ has capacity $C_n$ and state of charge $SOC_n$, both of which are known for all cells. The total amp-hours available $AH_n$ from cell $n$ is then $C_n \times SOC_n$ and the total amp-hours of charge $\min(AC_n)$ that each cell can accept is $C_n \times (1 - SOC_n)$. The pack can be defined in the fully charged state when $AC_n$ for one or more values of $n$ equals zero, and similarly the fully discharged state occurs when $AH_n = 0$. The total amp-hours available from the pack, therefore, is equal to the minimum value of $AH_n$ when $AC_n$ equals zero—that is, the available charge from the cell containing the smallest available charge (which is not necessarily the cell at the lowest SOC or the cell with the lowest capacity) when the pack is in the fully charged state.

At any given point in the charge/discharge cycle, the SOC can be calculated using the following method. The total amp-hours that can be discharged is $\min(AH_n)$ and the total amp-hours that can be charged is $\min(AC_n)$. The nominal capacity of the pack is equal to the sum of these two values. $C_{pack} = \min(AH_n) + \min(AC_n)$. $SOC_{pack}$, therefore, is equal to $\min(AH_n)/\min(AH_n) + \min(AC_n)$.

In many cases the differences in cell capacity and SOC are small, and therefore simplifications can be made. If the differences in capacity are negligible, then the available capacity is affected only by the state of balance. The effective pack capacity is reduced by the difference in state of charge of the most charged and least charged cells, multiplied by the nominal cell capacity.

For example, assume that there is a 3% difference in SOC between the most charged and least charged cells. The nominal cell capacity is 100 Ah. The pack reaches 0% SOC when $SOC_{min} = 0\%$ and $SOC_{max} = 3\%$, and the pack reaches 100% SOC when $SOC_{max} = 100\%$ and $SOC_{min} = 97\%$. The pack SOC can be computed as $SOC_{pack} = SOC_{min}/(100\% - (SOC_{max} - SOC_{min}))$.

The computational power required to run the full model on all cells may be too costly for some applications, especially with large numbers of cells. In this situation, a reduction of complexity is required to achieve acceptable performance with a smaller computational overhead. A number of scenarios are possible.

The least intensive solution is to run the model on an aggregate pack voltage or average cell voltage. This requires only a single instance of the model to be run at one time but has a number of disadvantages, particularly if the cells are not well matched. As discussed above in the full model implementation, the total available capacity is affected by the capacity and amp-hour mismatch between the individual cells. An average-voltage model will not fully capture the impact of these types of imbalance. Furthermore, with an imbalanced pack, the shape of the SOC-OCV curve will be distorted if an average voltage is used.

Another approach would be to use the median cell, defined as the cell that most represents the average condition of the pack, but without including the distorting effects of using information from outlying cells.

Incorporating more information, it is theoretically possible to determine the overall pack state of charge simply from knowing the SOC of two cells: the cell with the fewest amp-hours available to discharge and the cell with the fewest amp-hours available to charge. As previously discussed, these factors are the combined effect of both cell SOC and capacity and therefore the cells with minimum and maximum voltages do not necessarily represent the two critical cells for determining overall pack SOC. Further complicating the problem is the fact that these two critical cells are in many cases not the same cells throughout the life of the battery system—capacities and state of balance are transient metrics that change over time. A final complicating issue is that these values tend to change relatively slowly and require collection of a significant amount of data to update them.

## 19.6 Balancing

The available energy in each cell is changed by cell balancing activities. Unlike the overall battery current, the current that flows due to cell balancing is not identical for all cells and in many cases is not measured by current sensors. Because many battery models are designed to be tolerant of small errors in process and measurement, if the balancing current is small compared to the battery's capacity, the battery model may automatically compensate for the effects of cell balancing. If the balancing current is larger, then errors in SOC and other calculated values may persist and an active compensation method should be employed. Assuming that individual current sensors are not available for each cell, the balancing current needs to be calculated based on the type of balancing circuit used and the cell voltage. This current can be integrated and deducted from the amp-hour availability of the cells being discharged. In the case of active balancing circuits, this calculation can become more complex because energy can be moved between cells.

## 19.7 Temperature Impact on State of Charge Estimation

The dependency of available capacity on temperature has been discussed. Even if it is assumed that the temperature of all cells is known and an exact capacity value can be derived for each as a function of temperature, this alone is not enough to provide accurate run-time prediction, especially in situations using battery thermal management.

If the temperature is expected to change during battery operation, it will be the temperature at the end of the discharge that will define the total available capacity for that charge cycle. If the temperature is not being controlled closely by a thermal management system (which is often the case, in the interest of energy conservation), the temperature at the end of the discharge must be predicted to accurate estimate available capacity.

The battery's temperature at a time $\Delta t$ in the future will be the current temperature plus the integral of the net heat flow between now and $(t + \Delta t)$, multiplied by the battery's heat capacity.

$$T(t + \Delta t) = T(t) + C_p \int \dot{Q} dt$$

Total heat flow into the battery cells can usually be divided into three components: internal heat generation due to battery self-heating, passive heat transfer between battery cells and their environment, and (if applicable) forced heat transfer due to active thermal management. Internal heat generation can be calculated based on the discharge current profile and the ohmic resistance of the battery cells. Passive heat transfer between the cells and the surroundings can be estimated with a lumped parameter model defining the heat capacity of various components and thermal resistance of the interface between each.

In many cases the control strategy for the thermal management system will be known and the commanded heat flow is a function of battery temperature. The closer the battery is to the end of the discharge, the more accurate the current temperature can predict the final temperature since the total heat flow that is possible during the remaining discharge time is reduced.

Achieving a useful prediction of the end-of-discharge temperature requires assumptions to be made about the future rate of discharge so that the total estimated remaining run time and self-heating can be estimated. For some applications this is straightforward as the current profile is known in advance and in others it is harder to predict.

# 20

# Safety

## 20.1 Functional Safety

The concept of functional safety is defined by a number of standards such as IEC 61508 and ISO 26262 (specific to the automotive industry). These standards encompass a set of requirements and best practices necessary to define, design, implement, test, and deploy systems that carry out safety-critical functions. Functional safety is defined as "the part of the overall safety of a system or piece of equipment that depends on the system or equipment operating correctly in response to its inputs." [1] A battery management system must respond correctly to its inputs to ensure battery safety.

Large-format lithium-ion battery systems, due to the hazardous nature of many battery conditions, often imply that the battery management system must be developed to comply with an appropriate functional safety standard.

## 20.2 Hazard Analysis

In the concept of the battery management system, hazard analysis is the process of determining the possible dangers that could result due to a failure of the system to perform one or more of its functions correctly.

Standard hazards that should nearly always be considered are:

- Failure to prevent overcharge;
- Failure to prevent overdischarge;

---

1. From ISO26262.

- Failure to prevent overcurrent;
- Failure to prevent operation at extreme temperatures;
- Failure to respond to a disconnect command;
- Failure to interpret an open interlock signal;
- Failure to detect and respond to a ground fault.

Mission-critical battery systems without a fail-safe option could also include hazards such as

- Overestimation of battery state of charge;
- Overestimation of battery state of health;
- Underestimation of battery capability or capacity;
- Failure to disconnect the battery if its operating limits are exceeded.

For each hazard, a level of risk is assigned. ISO 26262 and IEC 61508 describe three components of risk: severity, occurrence, and controllability.

*Severity* describes the maximum possible level of danger, damage, or injury that could result based upon the occurrence of the hazard. It is important to understand that hazard severity depends not only upon the battery management system, but upon the battery system as a whole as well as the application and even possibly the specific action or mode of operation. For example, a battery-powered aircraft that loses power while taxiing is a much different situation than if power is lost during flight. The severity also changes depending on whether the aircraft is manned or unmanned. Consider that the battery cell and the battery management system could be the same for all of the above applications.

In the context of battery management systems, the description of the hazard should be very specific to address the potential failure or fault of the battery management system, not of the complete battery system or application. For example, battery overdischarge, while an undesirable condition, may occur due to self-discharge during long periods of disconnection without any defect in the battery management system at all. The corresponding battery management system function could be to detect overcharge and prevent operation of an overdischarged battery. Therefore, the hazard could be better described as "failure to prevent operation of an overdischarged battery."

In the case of exposure to external sources of heat that raise the battery system to a high temperature where thermal runaway occurs, again, there is no failure of the battery management system associated with this dangerous condition. While a battery system should always consider appropriate levels of protection against external heating and fire, relying on the battery management system to prevent overtemperature and thermal runaway in extreme conditions is not appropriate. However, it is reasonable to assume that a functional battery management system is required to accurately measure battery temperature and disconnect the battery from the load or otherwise restrict current flow if the battery begins to overheat. The failure to do so can be considered a battery management system hazard.

The battery management system is often expected to respond to safety signals or interlock circuits that indicate that the system has been placed into an unsafe state. Incorrect interpretation of these signals could result in the activation of a battery system during repair or maintenance causing an electric shock hazard.

*Occurrence* describes the frequency or probability of conditions in which the hazard could occur. For example, charging, and therefore overcharging, can only occur when the battery is connected to a source of energy. In the above example batteries installed in a climate-controlled building in moderate climates are unlikely to ever experience extreme temperatures. Occurrence in this context describes the occurrence of the prerequisite conditions for the hazard, not of the hazard itself, which is normally much lower. For example, overcharging could occur any time the battery is charged; charging is normally a very frequent condition, whereas overcharge should be exceedingly rare. In contrast, the maintenance of a battery system occurs much less frequently, therefore a hazard that could cause injury to maintenance personnel might receive a lower occurance rating.

*Controllability* is a measure of the degree to which the maximum hazard level described by the severity rating could possibly be avoided. Consider a medical device that could be either implanted or external. The maximum level of personal injury associated with a battery fire could be identical in both cases; however, with the external device, there is a much higher probability that this severe result can be avoided.

Battery fires, explosions, or other thermal events are determined to be possible outcome of a number of battery management system hazards. Establishing an upper bound for the severity of fire hazards is often very challenging unless a limit on the total energy released can be established. The total energy released in a fire may exceed the potential thermal release of the batteries themselves if additional sources of fuel are present. It may be possible for some small systems to prove that the maximum injury or damage possible is very limited, but under many conditions fires can spread and cause significant injury and damage. The occurrence of some battery management system hazards that could lead to fire also varies. "Failure to prevent overcharge" hazards are only possible when the battery system is charged, but many systems in grid storage and electric vehicles capable of regenerative braking are potentially charged at any time during normal battery operation. Overcurrent is another common cause of thermal events, but passive overcurrent devices (the failure of which is not strictly a battery management system hazard) are often employed to prevent overcurrents. Another possible cause is operation at elevated temperatures or failure to engage thermal management or fire suppression.

Another possible result of many hazardous conditions is electric shock. The potential maximum severity of an electric shock depends upon the energy content of the electricity source (IEC 60479 is one possible method of quantifying electric shock severity). Receiving a shock in many cases requires that either the battery is being serviced and protective covers or devices have been removed, or the battery has been somehow damaged (a motor vehicle crash is one possible way). The exposure ratings for shock hazards should reflect this. Controllability of shock hazards can be improved through the use of proper grounding schemes, touch-safe components such as connectors, multiple layers of insulation, and keeping Y-capacitances to a minimum. A battery management system that fails to detect an open interlock circuit potentially creates a hazardous condition that may lead to a shock hazard.

Detection of ground or isolation faults is a requirement for many systems. The maximum severity of an undetected ground fault will vary with the system design. Ground faults may lead to shock hazards or battery short circuits if additional faults develop or accidental contact is made between battery components and earth ground during servicing. The occurrence of a hazardous condition requires not only the failure of the battery management system to detect the ground fault, but the ground fault must actually occur.

A hazard rating is assigned to the hazard based upon these three rankings. It is important that the hazard rating be assigned appropriately as high hazard ratings will be excessively demanding in the design phase. When assigning battery hazard rankings, knowledge of the complete battery system, as well as the load application, is required.

With large-format lithium-ion systems, it may be helpful to distinguish hazards associated with individual cells from those associated with all cells or large number of cells. For example, it may be useful to consider the following as three separate hazards, each with different ratings for severity and controllability:

- Failure to prevent overcharge of a single cell;
- Failure to prevent overcharge of an entire module;

- Failure to prevent overcharge of the entire battery pack.

Understanding the battery cell behavior is necessary to improve the specificity of the hazard description. For example, if a margin of safety of 10% has been clearly established during overcharge testing, a good overcharge hazard could be "failure to prevent overcharge by 10% or more of a single cell." Overcharges of less than 10% could be a different hazard of lower rating, or deemed to be not to be hazardous at all (provided that subsequent cycling of the cells could be prevented).

A number of dangers associated with large-format lithium ion battery systems have been discussed as not strictly being considered battery management system-specific hazards. This does not imply that extreme care and an appropriate level of skill and testing are not required. The purpose of this discussion is to ensure that the appropriate safety analysis is conducted on each portion of the complete battery system.

The expense, complexity, and care required to address hazards with high ratings for severity, occurrence, and controllability are considerable. A good understanding of the battery system and application will ensure that hazard ratings are appropriate and do not lead to undue cost in battery management system design.

## 20.3 Safety Goals

For each hazard, a safety goal is established. The safety goal is to prevent or control the occurrence of the hazard. In functional safety frameworks each safety goal is assigned a safety integrity level, which defined the level of precaution and care which must be taken to prevent the hazard from occurring.

In automotive systems, according to ISO 26262, four Automotive Safety Integrity Levels (ASILs) are defined from ASIL A to ASIL D. Hazards that are assigned an ASIL D rating require the highest level of care and rigor to prevent them from occurring, as they will almost certainly cause loss of life or serious injury.

IEC 60518 defines similar Safety Integrity Levels (SILs) from SIL 1 to SIL 3.

For a given safety integrity level, the applicable functional safety standard lists the activities that must be performed to meet the standard. Some of the activities are related to the engineering process while others relate to the product itself. A number of activities have well-defined metrics to determine if the standard is met whereas others are more esoteric. Some examples of requirements for high-level safety goals can include:

- Demonstrate that the likelihood of a random failure in the control system leading to violation of the safety goal is extremely low (less than one failure per $10^8$ operating hours).
- Perform formal deductive and inductive failure analysis for possible ways in which safety goals could be violated.
- Use only qualified tools for software development (such as compilers) to prevent error-free source code from being translated incorrectly to assembly code that contains a vulnerability.
- Demonstrate traceability among system, subsystem, and component requirements and validation cases.
- Use software language subsets such as MISRA that aim to reduce the risk of certain types of errors.

## 20.4 Safety Concepts and Strategies

ISO 26262 and IEC 61508 discuss the idea of a *safety concept*. The safety concept is the method used to prevent hazards from occurring and to accomplish the safety goal.

Redundant measurement and control are an effective method to achieve safety goals in many control systems and battery management systems are no exception. In modern embedded systems, due to the complexities of hardware and software, it should always be assumed that latent defects may exist and therefore there are an unknown (and very high) number of failure modes with a nonzero chance of occurrence. A redundant system aims to reduce the risks associated with these types of failures by replicating multiple instances of the susceptible system, of either identical or different design.

Many of the available stack monitoring chipset solutions support these types of redundant architectures that can produce a lower-cost implementation than a system using only single measurement and control path with the extremely high reliability needed to prevent high-level battery hazards.

## 20.5 Reference Design for Safety

The need for integrated system safety has been discussed many times as well as specific requirements for meeting standards and an approach to include safety in the implementation of hardware, software, and mechanical design. This section will discuss a possible solution to address many of the needs of a modern, large-format battery management system.

The high integrity to prevent dangerous failure modes and safety goal violation is met through the use of a redundant architecture. The system uses a secondary signal and control path by which critical failure modes such as overcharge or a battery management system defect that prevents contactors from opening are avoided.

A low-cost approach to achieving this is discussed here.

A redundant stack monitoring architecture is chosen, using primary and secondary monitoring ICs. The secondary monitor IC provides only a small number (usually one or two) of digital outputs in the event of an overvoltage, undervoltage, or overtemperature fault. These signals are passed from north to south using the IC's internal level-shifting interface. In the case of a distributed design, this signal is converted to a chassis-referenced signal (referenced to earth potential) inside each slave module; for a monolithic design, only a single isolation barrier is needed at the lowest potential device. This signal uses current-mode signaling or a PWM output so that the master device can differentiate between a disconnected signal and a true battery fault.

A secondary, lower-power, low-cost microcontroller is used to provide secondary monitoring functionality. This microcontroller is responsible only for safety, and therefore its firmware is kept as simple as possible. The safety microcontroller and the main processor communicate by way of $I^2C$ or SPI. A minimum of resistive isolation prevents conducted transients from one processor affecting the other via the $I^2C$/SPI interface. The power supplies for both microprocessors are of a diverse design, and ideally the safety protection microcontroller uses a simpler power supply architecture that may be less efficient but more robust to transients simplified schematics of master and slave devices are shown in Figure 20.1 and 20.2.

The two microcontrollers exchange a seed and key through the $I^2C$/SPI interface to ensure that each microcontroller is functioning correctly. Each of the two microcontrollers is able to cause a reset in the other.

The system contactors use a common high-side drive switch with incorporated short circuit protection and current sensing. Each individual contactor is controlled by a single low-side driver also with incorporated short circuit and overload protection. The high-side driver is controlled by the safety microcontroller, whereas the individual low-side drivers are controlled by the main microcontroller. This architecture requires both microprocessors to be operating correctly to cause both the high-side and low-side drivers to be enabled. The split power supply architecture and separate processors prevents a single power supply transient from creating a failure where all contactor drive circuits are activated.

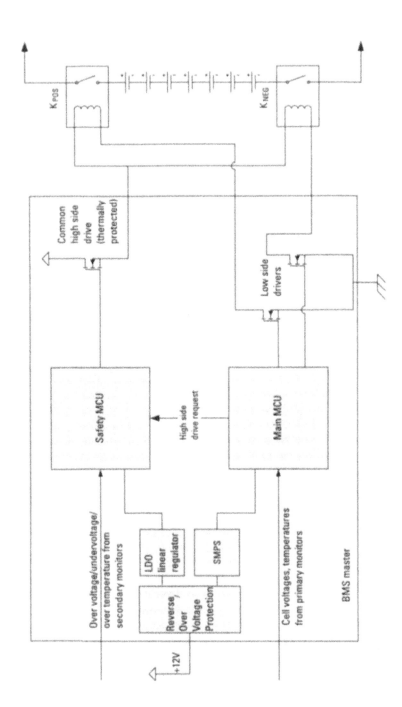

Figure 20.1 Reference design for master module.

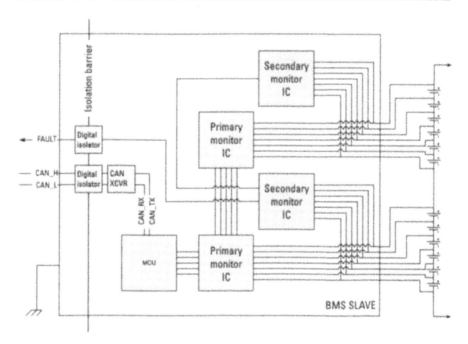

**Figure 20.2** Reference design for slave module.

# 21

# Data Collection

## 21.1 Lifetime Data Gathering

Battery management systems are often expected to gather and maintain information about the environment that the batteries have been exposed to during their lifetime. Suppliers of battery cells, systems, and battery management systems are all concerned with the dependency of battery longevity and performance on the operating history, and, therefore, storage of this type of data should be expected to be part of the battery management system requirements.

Simple scalar accumulators such as total time in service, amp-hour, and watt-hour throughput (which can be measured in both in charge and discharge) are commonly used to gauge battery usage. Counting cycles is also another common technique, but the definition of what constitutes a single cycle will often vary from one application to another as many applications experience partial cycles of varying depths of discharge and charge. For some applications, maximum and minimum current and SOC may be useful as well.

History of temperature extremes is also important to understand. Maximum and minimum temperature values are useful but do not give much useful information beyond the possible root cause of a system problem caused by extreme temperature exposure. Because many temperature effects are nonlinear, integration of temperature with respect to time also fails to adequately describe the thermal environment the battery has experienced.

A straightforward but effective strategy is to discretize the temperature space into a finite number of temperature ranges and to discretize the operating time into packets of time (see Figure 21.1). The operating history is then more accurately described showing the amount of time spent at high and low temperatures.

This approach can be extended to create a two-dimensional matrix of temperature and SOC (the two most significant environmental variables for battery degradation) (see Figure 21.2). In addition to operating time, the total amp-hour throughput in both charge and discharge could be captured at each condition.

Finally, the two-dimensional matrix can be extended into a third dimension to capture the battery current, which represents the third driving factor for degradation rates. For given ranges of rate, temperature, and SOC, a record of the total time spent at each condition gives a good picture of the total operating history of the battery that can be used to determine the impact of each of these factors on battery life.

Additionally, certain notable events may warrant the creation of a data log. Many applications require the recording of the system state in the event of a battery fault occurring. A snapshot of the battery voltages, currents, temperatures and state of charge as well as status of additional inputs and outputs is helpful in the diagnosis of the cause of the fault condition.

The opening of contactors under load is an event that usually can occur only a limited number of times while maintaining system safety (contactors are usually rated for a small number of full-current circuit interruptions). If this is

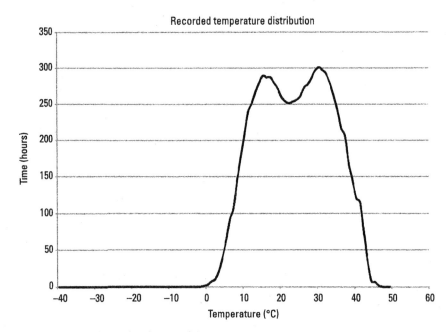

**Figure 21.1** One-dimensional array of time at temperature.

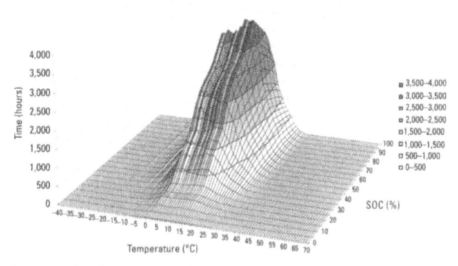

**Figure 21.2** Two-dimensional array of time at temperature and SOC.

a concern, the battery management system should maintain a record of these events and indicate the need for contactor replacement.

Because large-format lithium ion battery systems often have service lives of 5 to 20 years, depending upon the application, the ability to record and retain data should support data gathering over such a time span. This will drive the requirements for nonvolatile devices, which typically have a maximum number of write/erase cycles.

The method for extracting logged data from the battery management system will vary considerably depending upon the application. In automotive applications, fault codes (DTCs, or diagnostic trouble codes) are often associated with a "snapshot" of the system data at the time of the fault. Unified Diagnostic Services (UDS) specified in ISO 14229 details a number of techniques for transferring data to an external device over CAN. This method is commonly used in the automotive service and support sector to extract data from vehicles in the field. Other applications that are grid-tied will likely have access to an Internet connection (possibly through a number of intermediate devices) and data could be transmitted in real time. Server-type equipment such as a UPS battery system could allow for a direct PC connection via USB. The data recording system should be robust against data loss during unexpected disconnections of the battery system and host/control power. Small onboard energy storage in the form of a battery (for recording of events after power loss) or a capacitor (for orderly shutdown and saving of the last "known-good" data if a power failure occurs) helps improve the reliability of data storage.

# 22

## Robustness and Reliability

The design of a battery management system, especially for a safety-critical application, should be shown to be robust in the nominal cases through basic design calculations, and the worst-case performance can be analyzed for many circuits based on extremes of the expected values of the components. Software can be created according to best practices and analyzed with tools to show that it is free of certain types of defects. Testing can verify a number of systems over a wide range of operating conditions. However, with any modern embedded control system, there is the possibility that, through misuse, component failure outside the published limits, environmental exposure, aging, or other factors that are difficult to predict, the system will experience a fault and fail to perform some or all of its functions. The complexity of this type of systems makes the problem of predicting all of these possible failure modes impossible. As such, engineering battery management systems for reliability and robustness are critical to ensuring successful deployment of large-format battery systems.

Robustness, in the engineering sense, describes a system which is insensitive to various types of noise—noise being factors that may influence the system's performance that the designer cannot control. Good examples of engineering system noise present in most electronic systems include:

- Operating temperature;
- Electromagnetic fields;
- Age;
- Duty cycle;
- Manufacturing and component variations.

In the case of battery systems, these can translate into the following types of effects:

- *Initial capacity, impedance, and self-discharge variability between battery cells, packs and modules:* Algorithms should be tested with worst-case estimates of these parameters and verify that satisfactory performance is achieved.
- *Measurement bias for voltage, current, and temperature:* From unit to unit, some random variability will exist due to component variance. Worst-case analysis should exist to demonstrate that critical component tolerances are properly specified and adequate testing exists to ensure that inaccurate components will not lead to inadequate performance. In some cases, calibration or trimming of individual units may be achieve higher performance at lower costs than expensive, high-precision parts.
- *Duty cycle and application use:* The duty cycles imposed by all users of a particular design of battery system may vary widely. Aside from the effects on the battery cells themselves, the load profile may influence state of charge and state of health accuracy, balancing capability, and other core battery management system functions. The variability in the use cases and the impact of this variability on performance must be understood and the battery management system should demonstrate acceptable performance under all reasonable use profiles, even if they may lead to rapid battery degradation. Efforts should be made to define test cases corresponding to unusual operating conditions that may cause certain types of failures.

## 22.1 Failure Mode Analysis

Failure mode analysis is an important tool used in many disciplines for improving the robustness and safety of engineered systems. A comprehensive failure mode analysis should be conducted for any large-format lithium-ion battery system complete with its battery management system.

Battery management system and battery system FMEA activities must complement and reference each other. Ideally, a cross-functional group involved with the mechanical, thermal, hardware, and software designs at both the cell and pack levels should participate in the failure mode analysis for both the pack and the battery management system.

An FMEA requires that each potential failure mode be assigned severity, occurrence, and detection ratings. A number of standards, such as SAE J1739, are used to guide the assignment of these ratings.

A number of potential failure modes to be considered are discussed here along with potential mitigation strategies:

- *Incorrect cell voltage measurement:* Incorrect cell voltage measurement can render the battery system inoperative, potentially lead to undetected overcharge or overdischarge conditions, lead to inaccurate state of charge and state of health, and limit calculations that may result in battery operation outside the safe operating area. Redundant measurement circuits can help; secondary measurement methods with limited accuracy can prevent overcharge or overdischarge but may not provide enough information to ensure that more complex calculations are correct. Errors with voltage reference devices used to provide a reference for multiple cells may result in errors being introduced to all of these cells. Voltage references can be checked against another voltage source, perhaps of lower accuracy (internal logic level power supply, for example) to protect against large measurement errors (which may prevent overcharge or overdischarge, for example, but not against errors in state of charge calculation). The technique of comparing pack and/or module voltage to the sum of cell voltages has been previously discussed; this can be used to protect against gross errors in voltage measurement that can lead to inability to detect unsafe conditions.

  Important distinctions should be recognized as appropriate between faults involving a single cell, a single parallel group large numbers of cells, and all cells. For example, overcharging a single cell is likely to be less severe than overcharging all of the cells, but the differences depend upon the specifics of the pack design; if the pack is robust against propagation from a single cell thermal event, the severity of these events may be very different; but if a single cell experiencing thermal runaway can lead to a larger number of cells also having thermal events, both events may be equally severe. These different scenarios can have different root causes and therefore different occurrence ratings. This is one example of how the impact of overall battery system design impacts the analysis and design of the management system.

- *Incorrect temperature measurement:* The two most common failures in a thermistor measurement circuit are due to an open or short circuit failure in either the thermistor or the interconnects. Measurement circuits should be designed so that shorts and opens appear as implausible off-scale high and low readings and are not confused with actual plausible temperatures.

- *Incorrect current measurement:* The use of a solid state current sensor with a dead band at each end of the measurement range can provide some detection window if the current sensor has failed. Furthermore, it is known that the current is necessarily zero when contactors are open, allowing for a check of the zero point measurement.

  The output polarity of Hall effect current sensors is orientation dependent and if the sensor is installed incorrectly, battery charge current will appear as discharge current. This will lead to the detection of negative impedance and should be recognized as an error.

  A secondary sanity check can be performed in many applications by comparing the current as measured by the battery with the current measured by the load device. This can capture measurement errors as well as sensor failures. Due to potential differences in sampling rates and measurement accuracy, measuring the difference in measured amp-hours by both devices over a moving window and using a leaky-bucket strategy is recommended to prevent synchronization errors from causing erroneous fault detection. This can capture sensor failures, excessive noise, sensor inaccuracies, and reversed Hall effect current sensors that reverse charge and discharge currents.

  A final verification can be performed using the battery cell voltage measurements themselves. The battery voltage should vary according to the battery model during application of current. A battery varying rapidly in voltage without current is likely equipped with a defective current sensor. Rapid changes in measured battery impedance are possibly due to scaling errors in the current sensor.

  This same method can be used to detect "stuck" cell voltage measurements in which the cell voltage does not appear to change during the application of large charge and discharge currents (the cell appears to have no impedance and no internal dynamics). This can be the result of electronics errors with the measurement circuit (CMOS latch-up can cause this), slave device microprocessors failing to report a new value, or software that is not properly processing new data. If all electronics are operating correctly, this problem can be the symptom of damaged cell interconnects where a number of parallel cells are disconnected from the string but remain connected to the measurement circuit. In any case, these faults can be potentially extremely hazardous if excessively high or low cell voltages are ignored.

- *Communications loss:* A common scenario is the loss of the communication link between master and slave devices in a distributed system architecture. This may occur due to EMI, loss of connectivity, or failure of the various components in the signal chain. A "virtual" loss of communication may occur if the software in either the master or slave module fails in a way that passes old, stale data to the higher levels of software in the mater module. It is vital that this condition be recognized quickly and invalid data is not used for battery state estimation. Except in the most severe applications, the system should be robust against temporary losses in communications and avoid interrupting battery functionality.

## 22.2 Environmental Durability

Battery management systems should be subjected to a comprehensive test and validation plan designed to expose the system to the worst-case hazards associated with the environment in which the battery management system will be operated, transported, and stored.

Some of these hazards and the possible modes of failure are listed here:

- *Mechanical vibration:* Mechanical vibration can cause cracking of PCB substrates and IC packages, especially large packages. High and low temperatures may exacerbate these failures. Clearances may be reduced during mechanical vibration if enclosure surfaces experience large deflections at resonant modes. Wire harnesses and tabs/cell interconnects/busbars are prone to failure as well and may need to be included as part of a comprehensive suite of tests.
- *Mechanical shock:* This is similar to vibration, with special vulnerabilities for large/heavy components.
- *High temperature:* High temperature increases the rate of degradation of many components. Optoisolator degradation and dielectric strength can be affected. Temperature also significantly affects the accuracy performance of nearly all stack monitoring chipsets.
- *Humidity:* High humidity can create a number of conditions which reduce dielectric breakdown strength. The battery management system must be powered up during the humidity test to gain a full understanding of the effects of applying high voltage in a humid environment.

- *Drop:* A drop test should be performed to ensure that dropping of battery management system modules does not create latent damage that could go unnoticed.
- *EMC:* Voltage, current, and temperature measurement inaccuracies may occur at relatively low disturbance levels. These errors may be small relative to the quantities being measured but significant in their impact on the battery management system's calculation of battery parameters.
- *ESD:* Electrostatic discharge can damage input, processing and output devices and can reach the battery management system during handling, battery system assembly, service, or operation. Electric vehicle charging ports or user interface connections can be potentially accessed by users while the battery system is in operation and must meet "active" ESD requirements that can often be as high as 25 kV (as opposed to "handling" ESD which requires only tolerance of 4–8-kV discharge).

This test plan will usually consist of four phases and require a set of sample devices.

- *Initial performance characterization:* The devices will be subjected to a comprehensive evaluation of their performance. Critical parameters include cell and string voltage accuracy, power consumption from control power sources and high voltage stack, and dielectric strength. Note that the dielectric withstand test may cause damage to devices and may need to be performed on a special batch of devices to gather performance data about typical devices for comparing to post-exposure samples. Devices that do not meet nominal performance specifications should not be used for tests.
- *Simulated environmental exposure:* A batch of devices is exposed to a series of tests designed to represent a worst-case situation for the expected application. A given sample set may be used for more than one test, if the effects of the tests are expected to be cumulative. Many tests will require the devices to be powered up and operational with a simulated environment (i.e., battery pack).
- *Post-test characterization:* The devices are recharacterized to determine if any changes in performance occurred during the testing. Comparisons to the pretest values should be made. In the case of high-accuracy measurements for current and voltage, which require high precision and depend on temperature and other conditions, special test procedures are needed to ensure that systematic deviations are not mistaken for changes

in performance. Dielectric testing for working and breakdown voltage ratings should be performed on all post-test devices as well.

- *Detailed teardown and analysis:* Inspection should be performed of the devices to look for signs of potential weakness in the design. Dendrite growth, tin whiskering, and other potential short hazards should be part of the inspection. Coating and enclosure integrity are also critical.

## 22.3 Abuse Conditions

Although the battery management system is commonly thought of being present to protect the batteries from abuse, the battery management system itself may be subject to a number of abnormal conditions during operation. The battery management system is a possible failure point for fire and electric shock hazards.

The challenges associated with EMC and ESD mitigation have been discussed earlier. Battery management system components must also be robust against control power transients and supply disturbances.

It should be demonstrated through analysis, testing, or both, that extreme cell and pack voltages, caused by overcharge, overdischarge, or a combination of excessive impedance and current, do not cause a short circuit failure in the measuring circuits. Adequate safety margin should be maintained for all inputs between expected and maximum tolerable voltage levels. Consider the effects of connecting the cell voltage measurements in an incorrect fashion that could expose inputs to large voltage errors including reverse voltages, groups of cells out of order, and large voltage differentials across a single channel. In many cases it is not necessary to ensure that the battery management system will continue to operate after exposure to all of these events, but always ensure that breakdowns leading to short circuits do not occur. If it is necessary to retain functionality, consider a more distributed battery management system architecture that limits the voltage in each individual module; this will increase the likelihood that a particular input combination can be tolerated.

Failures in the battery system may lead to battery management system abuse. If a busbar, cell, or fuse fails to an open circuit while current is flowing through the battery stack, the current may be forced through sense harnesses and the battery management system module. Even small currents relative to the battery capability will be very large compared to the battery management system and interconnect and will almost certainly cause damage. Careful sizing of interconnects and components will ensure that the damage is graceful and does not lead to a thermal event.

## 22.4 Reliability Engineering

Reliability engineering brings an analytical approach to the often poorly understood field of quantitative failure analysis. A reliability engineer seeks to quantify, through statistical means, the likelihood of a particular defect or failure occurring with a view to ensuring that failure rates are sufficiently low to meet the demands of a particular application.

Reliability engineering is applicable to large-format battery systems as they are often used in applications which are extremely sensitive to battery failures. Acceptable failure rates can be as low as one failure per $10^8$ operating hours for many uses of lithium-ion batteries for functions that are required to demonstrate the highest level of functional safety (ASIL D or SIL 3). Components involved in these critical functions must be selected from vendors capable of providing the required reliability data.

The reliability analysis must include all of the components in the signal chain for a given function. Preventing a battery hazard usually requires correct operation of measurement, processing and control circuits, each of which contains many components.

# 23
# Best Practice

## 23.1 Engineering System Development

An appropriate model for system development should be applied to battery management system projects. The commonly referenced V-cycle for the development of engineering systems consists of definition of system, subsystem, and component requirements in a top-down manner to a detailed, structured process for implementation, followed by a bottom-up validation from component to system level, with traceability against the requirements established at each level.

The creation of the battery management system requirements will ultimately have its roots in the requirements of the overall battery system. The fundamental needs of a battery system are to supply and accept a given quantity of energy at a specific power rating, and to do so safely throughout the life of the system. The battery management system supports this set of goals to a large extent. The requirements developed for the battery management system are always cascaded from the needs of the next higher level engineering system and not developed in isolation without an understanding of the battery system's response.

The implementation of a battery management system requires good alignment of requirements at the solution level (battery and load requirements must align with each other), battery system level (accord between cell, pack, battery management system, and auxiliary component requirements), and battery management system level (including hardware, software, mechanical, and thermal integration). Processes should ensure that the influence of requirements at one level is always reflected in the lower levels. A collaborative approach among battery management system hardware and software engineers, system engineers, and battery cell and pack designers ensures complete coverage of all aspects of battery system development.

## 23.2 Industry Standards

There are a limited number of standards specifically for the development of battery management systems, but a number of standards authorities provide standards for integrated battery systems that in the case of lithium-ion systems will include the battery management system. These standards may or may not be legally required or binding depending on the application, location of use and installation. A detailed review of customer requirements and contacts, legal requirements for product liability, local and national electrical safety codes, product safety requirements, and/or motor vehicle safety standards is necessary to understand the full picture of required approvals for a battery system and battery management system in a particular market and application.

UL 1973 was developed for battery storage systems for rail applications but is now applicable to stationary energy storage for grid storage or integration of photovoltaic, wind, and other distributed generation resources. The standard specifies requirements for the battery pack protective circuit or the battery management system. This standard references a number of other standards for components and materials used in the construction of battery systems, notably including UL 1998 for safety-critical embedded software and UL 1642, which specifies requirements for lithium-ion cells.

SAE J2464 and J2929 discuss the safety of integrated automotive battery systems, including the battery management system, and cover mechanical abuse as well as a number of safety cases in which the battery management system response to a hazardous condition is involved. Federal Motor Vehicle Safety Standard (FMVSS) 305 specifies a number of requirements for isolation detection and high-voltage safety for electric and hybrid vehicles. The battery management system must often interface with charging equipment using a pilot line scheme described in SAE J1772. In the automotive market outside of North America, European standards regulate many of the same aspects, including UN/ECE Regulation 100, which covers requirements for numerous components of electric vehicles that can exceed 25 km/h, and IEC 62196, which describes the interface to charging stations (analogous to SAE J1772)

Many of these standards have similar requirements for performance in a variety of tests. Not all standards reference specific design elements or requirements; for example, the SAE standards are more heavily weighted on what the battery and battery management system must do rather than how it must do it. In contrast, the UL standards establish a cascading series of standards for various components and design attributes that span from cell level to system level.

## 23.3 Quality

The care taken in the analysis, development, and validation of the design of battery management systems must not be undone through defects introduced in manufacturing of the product. The final product is only as good as its ultimate realization, and therefore it is imperative that the manufacturing site subscribe to a quality management system that ensures that the battery management system that is produced meets all requirements.

Quality for a safety-critical product with high accuracy like a battery management system implies more than dimensional control and simple functional testing. While these are certainly necessary elements of appropriate quality control, the entire manufacturing process should be reviewed together with the design for possible areas where defects could occur with critical impact on the final product performance.

Areas of risk specific to the battery management system include:

- *Cleanliness and contamination:* Due to the presence of high voltages and possible arcing and fire hazards, the possibility of contaminating the final product with debris (especially if it is conductive) must be eliminated. This can affect equipment cleaning schedules, periodic automatic and manual inspections for foreign matter, and the use of appropriate covers or dunnage to prevent contamination. Die-cast parts should be carefully controlled to avoid conductive "flash" or overmolded material that could detach during product usage. Simple techniques like the orientation of components as they travel through a production process can avoid the accumulation of debris.

- *Coatings and dielectric materials:* The use of dielectric conformal coatings and encapsulants is common in the production of many types of electronic control modules, including the battery management system. If the use of these products is relied upon to achieve dielectric strength and insulation ratings, then the application must be strictly controlled to assure the applied coatings perform as intended. Appropriate methods include coupon testing of coated boards complete with dielectric testing, UV inspection to detect voids in the final coating film, and automated dispensing and application equipment. The use of liquid-applied coatings should be an area of careful review with the manufacturing organization.

- *PCB substrates:* PCB substrates must be supplied by a manufacturer familiar with the hazards associated with high-voltage electronics discussed in this book.

- *Components:* Components must be managed throughout their life cycle during the manufacturing process. Exposure to humid storage environments can degrade dielectric ratings for isolators and other components. Component bake-out is an appropriate control measure in some cases. Many low voltage control systems are not likely to experience safety-related failures due to improper component storage to the same degree as a large-format battery management system.

- *Component placement and alignment:* Components that are misplaced, including rotational and positional alignment and other defects such as "tombstoning," can create arc or shock hazards. Automated optical inspection (AOI) can be used to detect these types of defects but it is useful to concentrate inspection efforts on components attached to the high-voltage stack, which can create hazardous shorts if misplaced.

- *High current solder joints:* Some large-format battery management systems incorporate part of the high current path inside the electronic module. Large solder joints can crack while cooling and create high resistance connections and lead to heating.

- *High-current threaded fastener joints:* Improper torque increases connection resistance significantly at the interface between components using a bolted joint. High temperatures can result in this case.

# 24

# Future Developments

## 24.1 Subcell Modeling

Modeling tools are becoming available for the internal components of battery cells that can predict battery performance before a prototype cell is built. These tools can be used to begin the development of a battery management system before the cells are actually available, greatly speeding up the overall development cycle of battery systems. As processor performance for a given cost improves, more complicated models shift closer to being able to operate in real time. More complicated models will be able to be implemented in battery management systems of the future.

## 24.2 Adaptive Algorithms

One of the principal challenges facing battery management system development is the creation of a system that can handle more dramatic changes in cell behavior over the life of the battery system. Currently, a number of assumptions exist about the battery system during its life cycle, which reduces battery management system accuracy overtime. For example, the assumption about the SOC-OCV relationship being invariant as the battery ages does not hold if the capacity reduction of the two electrodes occurs at different rates, for example. Calendar and cycle life aging may lead to different types of effects. The model selected at the beginning of life may neglect effects that become important later in life making SOC and SOH estimation incorrect. As understanding of lithium-ion battery aging improves and battery systems are placed in longer service lives, models that account for changes in the battery performance more accurately will become commonplace.

## 24.3 Advanced Safety

The type of functional safety standards common to the aerospace and automotive industries are gaining increased acceptance in other areas of commercial, industrial, and even consumer products. As an example, UL is now specifying requirements for functional safety and applying regulations to lithium-ion battery systems. National electrical codes throughout the world will undoubtedly follow suit and regulations will become more complicated as battery systems are more widely installed.

## 24.4 System Integration

As battery systems become more commonplace and component volumes become higher, the battery management system will experience consolidation of functionality with other electronic control devices throughout the system.

A multilayered architecture can be envisioned that gives the battery system vendor the responsibility for basic safety and measurement, accurate SOC and SOH estimation, power limits and other battery-specific functions, with an application space in the same processor for high-level functionality that could change from one application to another.

# Endnotes

Plett, G., "Extended Kalman Filtering for Battery Management Systems of LiPB-Based HEV Battery Packs," *Journal of Power Sources*, Vol. 134, 2004, pp. 252–261

Prasad, G., and C. Rahn, "Development of a First Principles Equivalent Circuit Model for a Lithium Ion Battery," *ASME Dynamic Systems and Control Conference*, Ft. Lauderdale, FL, 2012.

Saha, B., and K. Goeblel, "Modeling Li-Ion Battery Capacity Depletion in a Particle Filtering Framework," *Annual Conference of the Prognostics and Health Management Society*, 2009.

Tröltzsch, U., P. Büschel, and O. Kanoun, *Lecture Notes on Impedance Spectroscopy: Measurement, Modeling and Applications*, Volume 1, CRC Press, 2011.

# About the Author

Phillip J. Weicker has spent more than 10 years as a pioneer in the area of electric vehicle propulsion and energy storage technology. He has played a leading role in battery system and BMS development with EnergyCS, Coda Automotive, and other organizations bringing electric vehicles and renewable energy systems to market. He specializes in the areas of systems engineering, simulation and modeling, technology commercialization, and safety. He has a number of patents pending in the field of electric vehicle technology. Mr. Weicker holds a bachelor's degree in electrical engineering from McMaster University in Hamilton, Ontario, Canada, and a master's degree in computational electromagnetics from McGill University, Montréal, Québec, Canada. He is a registered professional engineer in the state of California.

# Index

Balancing, 183–198
    charge transfer, 187–193
        flying capacitor, 188–190
        inductive, 190–192
        transformer, 193
    dissipative, 193–197
        thermal management, 195
    faults, 197–198
    sleep, 196
Battery models
    Approximation of nonlinear elements, 155–156
    Constant phase element, 154
    Doyle-Fuller-Newman model, 158
    Efficiency, coulombic, 153
    Hysteresis, 151–153
    Physics based, 158–160
    Randles cell, 150–151
    RC-equivalent, 149–150
    Self-discharge, 157
    Single particle model, 158–159
    State-of-charge dependent, 148
    Thevenin Equivalent, 146
    Warburg Impedance, 155
Battery pack
    architecture, 66–67

Charging, 111–114
    CC-CV Method, 111–112
    Constant current method, 113–114
    Target Voltage Method, 112–113

Capacity, typical, 45
    fade, 218
    rate-dependence, 200
Communication, 133–143
    CAN, 136–137
    Data integrity, 141–143
    Ethernet, 137–138
    FlexRay, 138
    $I^2C$, 134
    LIN, 136
    Modbus, 138
    Network Design, 138–142
    RS-232/485, 134–135
    SPI, 134
Connection, BMS to battery cells, 63–65
Contactor
    typical closing sequence, 55
Control
    contactor, 95–109
        chatter detection, 102–104
        economization, 104–105
        fault detection, 106–109
        precharge, 97–99
        topologies, 99–101, 105–106
        transients, 101–102
Creepage and clearance, 121–123

Data collection, 275–277
Distributed architecture, 61

Electromagnetic compatibility, 252–253
Electrostatic discharge, 128–130
Electrochemical impedance spectroscopy, 169–170

Fault detection, 231–239
   broken wire detection, 233, 235
   capacity loss, 238
   imbalance, 236
   lithium plating, 237
   internal short circuit, 237
   overcharge/overvoltage, 231–235
   overcurrent, 235–236
   overtemperature, 235
   self-discharge, 236
   venting, 237–238
FMEA (failure modes effects analysis), 280–282
   abuse, 285
   best practices, 287–290
   durability, 282–285
Future developments, 291–292

Hardware
   BMS ICS, 243–248
   circuit design, 250–252
   connectors, 241,249
   encapsulants, 241
   manufacturing, 254–255, 290
   microprocessors, 238-239
High voltage, 45–52, 119–134
   threshold, 45

Interface, 47
Hazards, 119–120, 123–134
Isolation, 125–128
   logical, 52
   physical, 51
   safety, 120–121
Isolation measurement, 130–132

Limit algorithms, 171–182
   cell voltage, 179–180
   fault conditions, 180
   overview, 171–173
   safe operating area, 173–174
   state of charge, 177–178
   temperature, 174–177

   violations, 181–182
Lithium ion batteries, 25–41
   operation and construction, 25–29
   chemistry, 30–33
   impact on BMS, 33–35
   safety, 35–38
   longevity, 38–39
   performance, 39–40
   system integration, 40–41

Measurement
   cell voltage, 71–77
      impact on SOC, 74–75
      redundancy, 74–76
      errors, 76–77
   current, 77–85
      accuracy, 84–85
      sensors, 78–84
         Hall effect, 81–84
         magnetostrictive, 84
         Shunt, 79–81
      signal chain, 78
   floating, 124
   inputs, 53
   interlocks, 92-93
   synchronization, 86
   temperature, 87–92
      accuracy, 90–91
      diodes, 92
      location, 90–91
      thermistors, 88–89
      thermocouples, 88
   uncertainty, 92
Modularity, 59–62
Monolithic architecture, 59

Operational Modes, 115–117

Parameters, 165
   identification
      brute-force, 165–166
      Kalman filtering, 168
      online, 166
      recursive least squares, 168
Power supply
   architecture, 253–254
   control power, 68–69
   measurement circuit, 67–68

Safety
functional, 265–270
    goals, 269–270
    hazard analysis, 265–269
    reference design, 270–273
Semi-distributed architecture, 61
Scalability, 65–66
Software, 69–70, 257–264
    architecture, 69–70
    analysis, 259–260
    battery model implementation, 262–264
    safety critical, 258–259
Standards, 50, 288–289
State of charge
    definition, 201–202
    estimation, 199–212
        Coulomb counting, 203–204
        Kalman filtering, 206–211
        OCV lookup, 205
        temperature compensation, 206
        voltage-based correction, 204–205
    impact on battery limits, 177–178
    relationship to OCV, 167
    Typical accuracy, 55

State of health
    capacity measurement, 223–225
    definition, 213–215
    degradation mechanisms, 215–216
    impedance measurement, 219–223
    parameter estimation, 226–227
    particle filters, 227–229
    predictive models, 216–219
    remaining useful life, 227
    self-discharge, 226
    Typical Accuracy, 55
State-space systems, 161–163

Thermal Management, 114–115

Variation, 47

Y-Capacitance, 125

CPSIA information can be obtained
at www.ICGtesting.com
Printed in the USA
BVOW11*2201020218
506838BV00008B/238/P